Civil Engineering Contract Administration

To Valerie

Civil Engineering Contract Administration

SECOND EDITION

A. V. Atkinson

Stanley Thornes (Publishers) Ltd

First edition published in 1985 by Hutchinson & Co.
(Publishers) Ltd

ISBN 0 09 159981 4

Second edition published in 1992 by

Stanley Thornes (Publishers) Ltd
Old Station Drive
Leckhampton
CHELTENHAM GL53 0DN
England

British Library Cataloguing in Publication Data
Atkinson, A.V.
Civil engineering contract administration.
– 2nd ed.
I. Title
624.068

ISBN 0-7487-1521-5

Photoset in Linotron Plantin with Eras
by Northern Phototypesetting Co. Ltd, Bolton
Printed and bound in Great Britain at The Bath Press, Avon

Contents

Acknowledgements

I am indebted to the Institution of Civil Engineers for permission to reproduce material from the following:

Conditions of Contract (sixth edition)
Guidance Note 2A: Functions of the Engineer under the ICE Conditions of Contract
Conciliation Procedure (1988)
Arbitration Procedure (England and Wales) (1983)
Civil Engineering Standard Method of Measurement (second edition)

I would also like to express my thanks to all who have given me assistance in the production of this book and, once again, to the following in particular, whose advice and help have been invaluable: Margaret Hollins, LLB, LLM, Solicitor; Derek 'Dick' Camp, RICS; Marion Densley; and Victor Towner.

Introduction

In the introduction to the first edition of *Civil Engineering Contract Administration* I claimed that there were many books to be found dealing with the administration of construction projects that suited the needs of building students, but relatively few that suited the needs of their civil engineering counterparts. That situation, as I see it, remains unchanged, which, together with a proliferation of new and revised standard forms of contract, new and amended legislation and, particularly, the revision of the Institution of Civil Engineers Conditions of Contract around which the book was centred, has inspired this second edition. The new (1991) sixth edition of the *ICE Conditions of Contract* has introduced numerous changes, many of which are minor in the interests of clarity, whereas others are comparatively important, resulting in extensive rewriting of various passages of the book.

My aim remains, as before, to help redress the balance, although much of the general text is equally applicable to both building and civil engineering contracts; to provide basic information generally for those who are new to civil engineering contracts or those who have previously been concerned with them only in a technical capacity and, more particularly, for those studying civils, highways, structures and quantity surveying higher national certificate units. It should be noted, however, that in respect of the latter, only the contract requirements relating to tendering, variations, measurement, certification and payment, are dealt with, and not the measurement process itself.

The principal objectives are to introduce the more significant administrative factors associated with civil engineering projects, and to establish the main contractual obligations placed on the various parties – the underlying theme being the value of effective contract administration. This book, therefore, is intended to be taken as a series of signposts directing towards smooth administration of civil engineering works and away from litigation, with the subject matter being related specifically to the sixth edition of the *Institution of Civil Engineers Conditions of Contract*.

The following pages are intended, therefore, to shed light on the preparation, execution, maintenance and payment of civil engineering contracts, which should be carried out in accordance with various established procedures, conditions and codes; the responsibilities and functions of the parties involved in the construction process and their interrelationship; the various standard forms and types of contracts; tenders and tendering; and the problems associated with safety, claims, disputes, bankruptcy and determination of the contractor's employment in particular. A general approach to the subject has been adopted, recognizing that the needs of those employed on the contractor's side of the industry vary somewhat from those of the engineer.

In addition to complying with the procedures and conditions, the execution of civil engineering contracts must conform with the broader requirements of the law: a detailed and complex subject into which a person such as myself, who is not a lawyer, must venture with extreme caution. I am especially indebted, therefore, to Margaret Hollins for her unstinting assistance and advice, particularly when writing the first chapter, which sets the scene for those that follow, providing a brief explanation of the law and the formation of contracts, sufficient for the purposes of this volume, but no more. For those interested in further study, suggested suitable reading material is: *Construction Law* (Sweet and Maxwell 1991) by J. Uff, *Law for the Builder* (Longman Scientific and Technical 1987) by Stephanie E. Owen, and *Osborn's Concise Law Dictionary* (Sweet and Maxwell 1983) by R. Bird.

Several chapters include brief references to, and the basic provisions of, a number of Acts of Parliament, as far as they affect civil engineering construction, for example Chapter 1 includes references to the Law of Property (Miscellaneous Provisions) Act 1989, Chapter 5 includes the requirements of the Occupiers Liability Acts, and the Health and Safety at Work etc. Act 1974; while Chapter 6 includes certain details of Acts relating to the sale of goods and the supply of services, etc.

All chapters, other than Chapter 1, make reference to relevant clauses in the *ICE Conditions of Contract*. Where clauses have not been dealt with in the general text they are summarized in Chapter 13 – Synopsis of the ICE Conditions of Contract.

A. V. A.

1 Law and the formation of contracts

Law may be described as the rules of a community or state to control the conduct of people in that community or state in respect of their private and business relationships and their relationship with the state itself. Law falls, therefore, into two principal divisions: civil law and criminal law.

Civil law

Civil law comes from various sources: Acts of Parliament (also known as legislation, or statute law), delegated legislation and case law. Case law is also known as common law, case precedent and judicial precedent and arises because not all the law is written down in Acts of Parliament or subordinate legislation, but may have been derived from court decisions in previous cases, when judges state their reasons for the judgements made. Those reasons provide the basis for future decisions: they set a precedent. The underlying theme of precedent is that all cases should be treated the same way as previous cases of a like nature. This has resulted in a series of decisions which has established sets of principles to suit given circumstances.

Civil law, from whatever source, deals with civil wrongs, where individuals or bodies bring cases against other individuals or bodies. If one party suffers a wrong, loss or injury resulting from the action of another party the wronged party can seek redress through the courts, by the issue of a writ. The redress will normally be compensation in the form of damages paid by the defendant to the wronged party.

Civil law is concerned with both general and arranged liability. It is subdivided, therefore, into two main classifications: tort and contract.

Tort

Tort can be defined as a civil wrong outside of contract. It may be difficult, in some instances, to draw a distinction between tort, other civil action or criminal action. Where this is the case, both a civil and criminal action may well be brought in respect of the same wrong. There are numerous torts, but the principal ones affecting the construction industry are those shown in Figure 1.1.

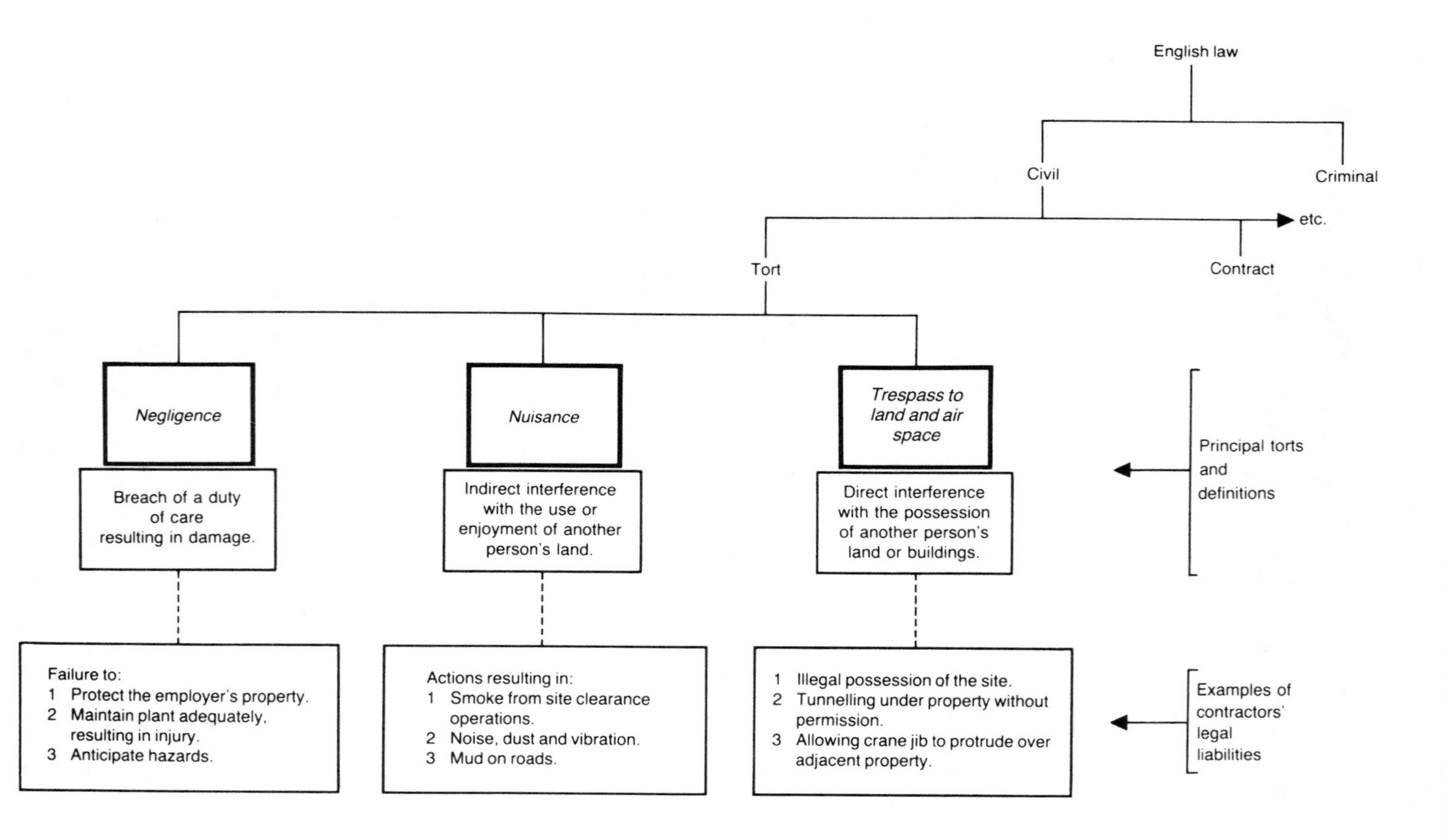

Figure 1.1 *Principal torts affecting the construction industry*

Contract

Contract can be described as liability arranged between individuals for their own convenience, as compared with liabilities imposed by law in a general way.

Civil actions are normally tried in the High Court but, where jurisdiction is generally restricted to £50,000, cases are heard in the County Court; and in minor actions, where the disputed sum does not exceed £1,000, in the Small Claims Court. This latter court provides for a relatively quick, inexpensive and easy resolution of disputes without the need for legal representation.

The High Court has three divisions:

1 *Queens Bench Division* – deals with most common law cases such as tort and contract.
2 *Chancery Division* – deals with cases concerning company and partnership disputes, bankruptcies, trusts, administration of estates, rectification of deeds and mortgages.
3 *Family Division* – deals with matrimonial cases, wardships, guardianships and property disputes between husband and wife.

Civil actions of a technical nature are often heard in the Official Referees Court before judges who specialize in trying such cases.

Generally, either party in cases brought before the High Court, County Court or Official Referees Court, has automatic right of appeal to the Court of Appeal. A further appeal may be brought to the House of Lords, but leave of appeal must be sought and is only given in extreme cases, on a point of law of general public importance.

It should be noted that construction contracts frequently provide for disputes between the parties to the contract to be dealt with by arbitration. Arbitration is dealt with in Chapter 9.

Criminal law

A crime can be defined as an act which is punishable by law. The criminal law's main function is punishment, as opposed to civil law's function of providing compensation or resolving disputes, and stems generally from enactments (Acts of Parliament) or from the common law.

Parliament may delegate some legislation to other bodies such as local authorities or to Government Ministers who prepare by-laws, rules and regulations. These are treated as being in the encompassing Act. Delegated legislation, or subordinate legislation as it is also known, plays an important part in the every day operation of construction work providing strict guide-lines and generally requiring more of those involved in the construction process than is required by common law. The Building Regulations is an example of delegated legislation, the authority for which is provided by the Public Health Acts.

Usually, the State in the form of the Crown Prosecution Service or the police will bring a prosecution against the alleged offender who, if found guilty, is punished by the State – with a fine, partial restriction of activities, for example a ban on driving for a limited period, or imprisonment.

Criminal actions are tried in either the Magistrates' Court or the Crown Court. The losing party in the Crown Court, wishing to appeal against the conviction or sentence, has to be given leave of appeal to the Court of Appeal, except where the appeal is on a point of law. Further appeal, to the House of Lords, may be brought, but this requires leave and is only given where a decision involves a point of law of general public importance.

Civil and criminal law

From the foregoing it can be seen that in general, civil engineering contracts are the concern of civil law and cases are usually heard in the High Court coming, in the main, before the Queens Bench Division. Alternatively, specialized cases will come before the Chancery Division. However, since no one person, business or contract operates in isolation, certain contractual situations may give rise to legal matters involving also the second division of law – criminal law. As an example, if a contractor contravenes, say, the Building Regulations and a third party is injured in the process, the contractor will be summoned by the State and will also have to pay damages to the injured party.

Since 1972 the United Kingdom has been subject to European Community Law, which mainly consists of delegated legislation made under the powers in the European Treaties. Community law restricts the power of the United Kingdom Parliament to make laws. It takes precedence over national laws and where there is a conflict between community and national law, the former prevails.

Contract law

A contract is a legally binding agreement between two parties, or an exchange of promises where one party agrees to provide something in return for something else from a second party.

Contracts do not have to be in writing, but can be oral or made by a gesture or, as at an auction, where the bidder makes an offer to the auctioneer, with a nod.

A contract can be as basic as buying a train ticket or making a shop purchase. In such cases there is nothing in writing other than a ticket or, maybe, a receipt for the cash paid. In other similar transactions a purchase may be made in, say, a supermarket where there is not even a spoken agreement. Such contracts are as binding as written contracts, but problems arise in the event of a dispute between the parties as it may be exceedingly difficult to establish exactly what was agreed and what the parties intended. As a consequence, engineering contracts are formalized in writing.

Contracts fall into two categories:

1 Simple
2 Speciality (i.e. made by deed)/Deeds.

Simple contracts
A simple contract requires consideration (see page 8) and, for the type of work with which we are concerned, will be 'under hand', that is, contain the signatures of the two parties. 'In writing' does not necessarily involve a formal contract but can be an exchange of letters, providing they include the essential elements of a contract which are described later in this chapter.

Simple contracts have a period of limitation of six years. This is a statutory period during which one party can start a legal action or arbitration proceedings, and will run from the date of the breach.

Although in general these contracts may be in any form, exceptionally some contracts cannot legally be oral. They fall into two groups:

1 Contracts which must be completely in writing, for example contracts for the sale of land and house purchase, bills of exchange, promissory notes, marine insurance contracts, share transfers and, in some instances, hire purchase contracts and consumer credit agreements.
2 Contracts which need not be completely in writing, but which must be 'evidenced in writing'. In these, all the material terms of the contract must be recorded on at least one note or memorandum. Contracts of guarantee fall into this group.

Speciality contracts (Deeds)
A deed is required under the Law of Property (Miscellaneous Provisions) Act 1989 to indicate clearly on its face that it is intended to be such, for example by use of the word 'deed'.

Deeds do not require consideration, although in many instances consideration will be present. They have a limitation period of twelve years, and statements of fact included in them cannot subsequently be called into question.

Formerly, deeds were documents signed, sealed and delivered by the parties to them. Following the Law of Property (Miscellaneous Provisions) Act 1989, deeds executed by individuals no longer require sealing: signatures in the presence of an attesting witness and delivery are sufficient. Deeds executed by a corporation, however, still require a seal (although a provision in the Companies Act 1989 may enable companies to execute a deed without such). Transactions which must be by deed include the transfer of legal estate in law, and legal mortgages of land.

Essential elements of a contract

There are a number of essential elements required for a valid contract to be formed:

1 Agreement between the parties.
2 Offer and acceptance of the offer.
3 Consideration to be provided by the parties.
4 Intention by the parties to create a legal relationship.
5 Genuine consent of the parties.
6 Legal capacity of the parties to enter a contract.
7 Legality of the contract.

Agreement

Agreement between the parties, regarding the purpose, rights and obligations which the contract will create is essential. A written agreement is not essential, but in practice it is desirable to put the agreement in writing since it will provide substantial evidence of the terms of the contract.

Although the parties to a contract are free to decide the terms (within the law) care should be taken that contracts are worded precisely, since the courts will implement contracts in accordance with those agreed terms. They are only concerned with what is written, not what is meant. If those terms are unjust, or if they become unjust, the courts have no power to change them, although the courts may be willing to imply a term to repair an obvious oversight of the parties. From this it will be seen that if a contractor contracts to carry out work at rates that will involve him in a loss, he must still do the work at those rates or face an action for breach of contract. He was not obliged to tender for the work, but having done so he must stand by his undertaking to do the work which was obtained at *his* rates.

Agreement is considered to have been reached when an offer made by one party is accepted by a second party.

Offer and acceptance

In forming a contract there must be an offer consisting of a definite promise from one party to the other of his willingness to be legally bound on specific terms, and an unconditional acceptance of those terms by the second party.

An offer may be withdrawn before it has been accepted, and will lapse after a 'reasonable' time if no time limit is imposed. Once an offer is rejected it is destroyed. Similarly, a different proposal destroys the original offer. If qualifications are attached to an acceptance it becomes a counter offer which can then be rejected or accepted by the original offeror. Alternatively, the offeror may attach conditions to his acceptance, thereby making a counter to the counter offer. In such a case, the acceptance must come from the original offeree. (See Figure 1.2.)

Advertisements, or exhibitions of goods for sale, are not usually offers but 'invitations to treat'. In principle, a shopkeeper does not have to accept an offer from a customer to buy since he is not bound to sell the articles on show. However, if it is clearly intended as an offer, for example an advertisement offering a reward, an advertisement may be deemed to be an offer.

It follows that a contractor's tender, which is his offer, does not have to be accepted by the client and that, strictly speaking, it is unnecessary for contractors to be informed when invitations to tender are sent out, that the client is not bound to accept the lowest tender, as often happens. In practice the client does not always accept the lowest and, on occasions, rejects them all.

The offer and acceptance must be communicated to have any effect and can be in writing, oral, or inferred. However, if the offer stipulates the manner of acceptance then the method stipulated, or an equally expeditious method, must be used. Thus, if an offer states that the acceptance shall be by 'return of post' acceptance takes effect from, and the contract is made at, the time of posting, regardless of loss or delay in the post.

The various stages in the formation of a civil engineering contract are shown in Figure 1.3. It should be noted that the offer comes from the contractor, not the employer offering the contractor the opportunity to compete for the work.

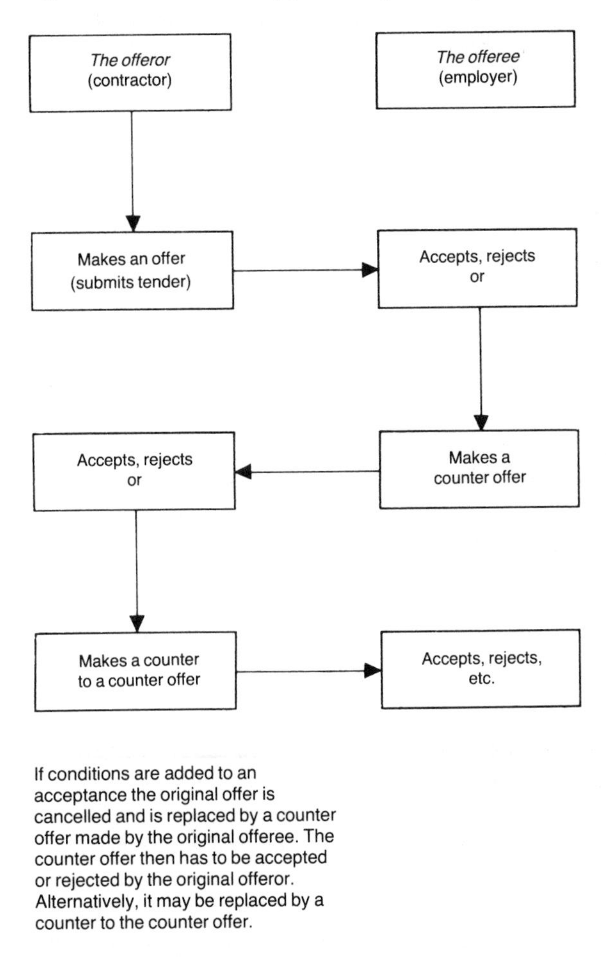

Figure 1.2 *Formation of a contract – offer and acceptance*

Consideration

Consideration is that which each party contributes to the contract. In other words, it is what the parties put into and get out of the contract.

Consideration must have some economic value, but the law is not concerned with the value being adequate. A contractor's low rate, therefore, may not be sufficient for him to make a profit, but if it is his offer, freely given, it will be deemed good consideration.

A contract under seal (speciality contract) does not require consideration.

Intention to create a legal relationship

The parties are required to form a contract enforceable at law.

In most business agreements there is a clear intention by the parties to create a legal relationship, and this is presumed to be the case by the courts. However, this presumption may be rebutted by a clear statement to that effect by, for example, including a statement that 'this agreement is not intended to be legally binding'.

Genuine consent of the parties

The agreement must be free from:

1 Misrepresentation
2 Mistake
3 Duress
4 Undue influence

Misrepresentation

Misrepresentation is a false statement of an existing or past fact which was made before, or at the time the contract was made, which misled the person to whom it was addressed and so induced that person to enter into a contract.

Misrepresentation falls into two categories:

1 *Fraudulent* (or negligent), where one party wilfully, recklessly or carelessly makes a false statement, or withholds material information. Here, the misled party can rescind the contract and sue for damages or, alternatively, take advantage of the contract and simply claim damages.
2 *Innocent*, where one party honestly makes a statement believing it to be true. Here, the misled party cannot claim damages, but can only rescind the contract. However, damages in lieu of rescission may be awarded by the courts if it is decided that rescission is too drastic a remedy.

Mistake

Errors of judgement made at the time the bargain was struck have to be borne by the parties. Mistakes of fact, however, may arise in contracts as to:

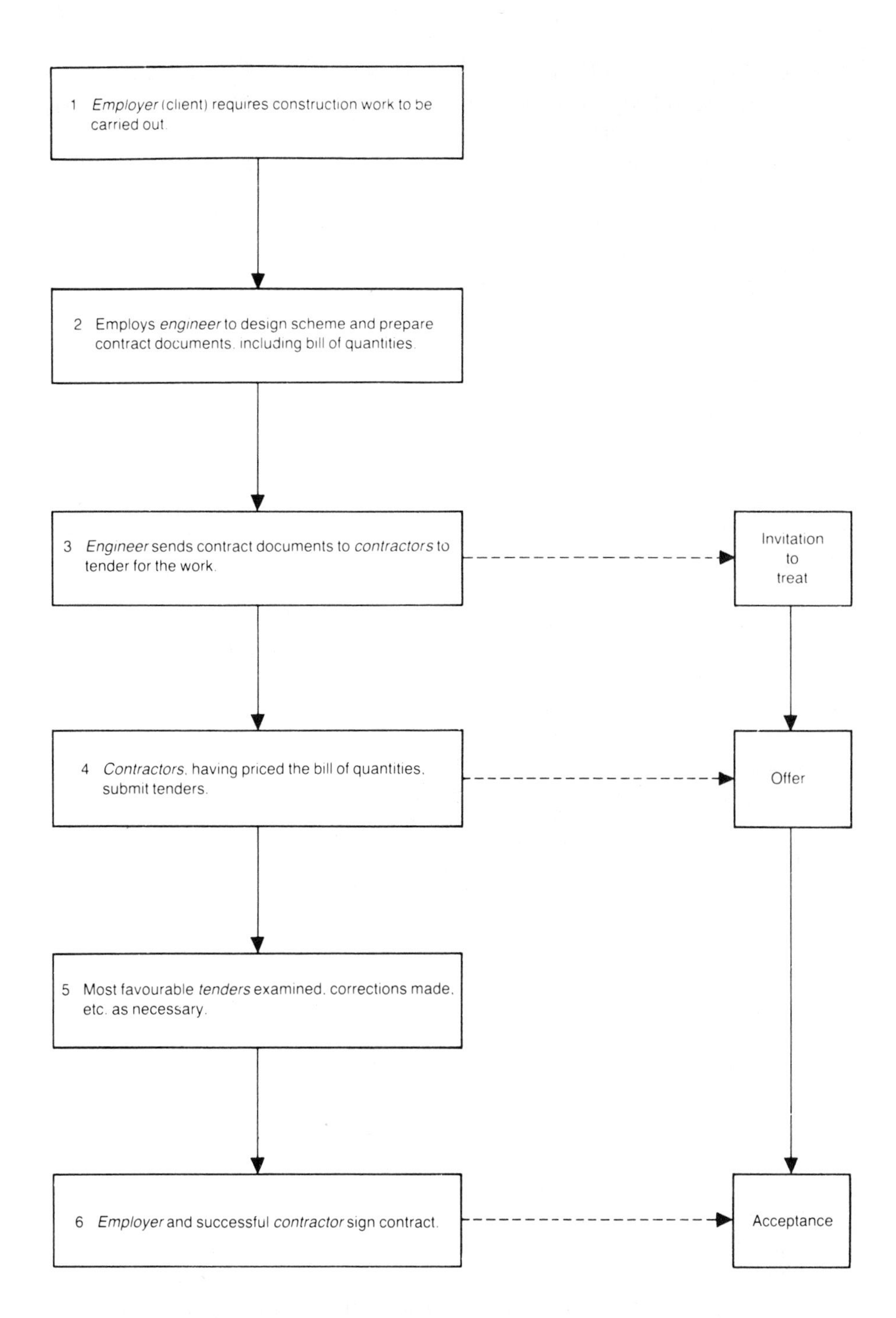

Figure 1.3 *Stages in the formation of a civil engineering contract*

1 *The identity of the other party.* For example, unknown to one party to a contract, the other party might have been taken over by a company that was unacceptable to the first party.
2 *The existence of the subject matter at the date of the contract.* For example, purchase of an article which, meanwhile, had been sold or destroyed without either party knowing about it.
3 *The quality of the subject matter.* For example, a painting was understood to be an 'old master' when in fact it is a cheap copy.

In 1 and 2 the mistake is fundamental to the agreement. As a result there is no agreement and, hence, no contract.

In 3 there is an agreement and, since 'let the buyer beware' is a general principle of contract, successful court action is unlikely.

Duress

It is generally recognized that a contract is voidable by a party if that party was forced to enter the contract as a result of actual or threatened violence to his person, but not to his goods. In other words, physical pressure or intimidation.

Undue influence

This is similar to duress, but the threats or influence here comprise any kind of improper pressure applied on a person to enter a contract. The word 'undue' is important, stressing that the contract was not made voluntarily.

Legal capacity

Certain parties are either not allowed to enter into a contract or are restricted in some way. These include persons under the age of eighteen, drunks, lunatics and convicts under sentence.

Corporate bodies, such as limited companies and public authorities, can only make contracts within the powers contained in their Memorandums of Association, and persons signing contracts on behalf of an organization should not do so unless they are specifically authorized to commit the organization.

In some contracts, the document itself states the names of the persons signing for the two parties and their positions. Such documents usually contain statements to the effect that those named persons are authorized to sign on behalf of the parties.

Legality

Contracts which are illegal, for example contracts which involve an act prohibited by statute, such as building without planning permission; or an agreement to commit a crime, or a tort or some immoral act, will not be enforced by the courts.

If the object of a contract becomes illegal after performance has begun the contract comes to an end and the parties are discharged from their obligations.

Legal contracts may, on occasions, be partially performed in an unlawful way.

If the illegal portion can be separated from the legal, the illegal portion only will be void, for example, although a contractor may contravene the Building Regulations and be summoned, the contract still stands and he will be paid for the work done.

Contract terms

The terms of a contract fall into two categories:

1 Express terms
2 Implied terms

Express terms
Express terms are statements of promises made by the parties, for example, the contractor's undertaking to carry out and complete the works by a specific date, and the employer's undertaking to make interim payments to the contractor.

Often, during the negotiating stage in the formation of a contract, statements are made by the parties which have to be written into, or appended to, the contract. Considerable care is required in preparing these statements due to their contractual effect.

It should be noted that an express term will normally take precedence over an implied term.

Implied terms
Implied terms are terms not expressly stated in a contract. Nevertheless, they are as legally binding as express terms.

Implied terms arise in three ways:

1 By custom or trade practice.
2 By statute, such as the Sale of Goods Act 1979, and the Supply of Goods and Services Act 1982.
3 By legal precedents established from previous cases.

Examples of implied terms include the understanding that goods and materials must be of good quality and fit for the purpose intended, and that the building owner will give possession of the site within a reasonable time.

Breach of contract

A breach of contract occurs when:

1 One party fails to fulfil his contractual obligations either totally or in part.
2 One party states that he will not perform the contract or, alternatively, puts himself in such a position that he will be unable to perform on the due date. This is known as 'anticipatory breach'.

In the first case, the contractual obligations may be discharged (brought to an end) if there is a breach of a condition (see below) leading to repudiation (refusal by one party to perform his obligations under the contract).

In the second case, the party affected by the breach has two options:

(a) To treat the contract as discharged from the date of the breach, or
(b) To wait until the due date of performance, in the hope that the other party will have a change of heart and perform. Choosing to wait may result in that party losing the right to recover if some intervening event occurs that will automatically discharge the contract, for example a war, or Government action making the contract illegal.

The terms of a contract legally fall into two categories:

1 Conditions
2 Warranties

Conditions

Conditions are important terms expressing matters basic to a contract. Failure to perform the requirements of a condition is a fundamental breach of an essential obligation giving the aggrieved party the right to:

1 End the contract and claim damages, or
2 Continue the contract and claim damages.

Warranties

Warranties are less important terms than conditions, and are subsidiary to the main purposes of the contract. In the event of a breach of a warranty, the aggrieved party has the right, only, to claim damages. An example of a warranty is where the employer undertakes to obtain any necessary planning permission for the works (see ICE Clause 26), or where a nominated subcontractor warrants that the work will be carried out to specific standards. Failure to meet the latter requirements could result in the employer obtaining damages.

This type of warranty should not be confused with a guarantee where for example a vendor undertakes to put right, free of charge within a stated period, any defect that might appear in a purchased article.

Privity of contract

Only the parties to a contract may normally benefit from a contract or be bound by its terms. Even if the contract has been made especially for a third party, that party cannot normally sue upon that contract. For example, a clause in an engineering contract entitling the employer to make direct payments to a subcontractor may be used by the employer, but cannot be enforced by the subcontractor, as he is not a party to the contract between the employer and the main contractor.

Generally, in civil engineering the employer has contracts with the engineer (his agent) and the main contractor, but there is no contract agreement between the employer and the subcontractors or between the engineer and the main contractor. The contractual relationships between the various parties to a contract are shown in Figure 1.4.

The contractual position is sometimes confused when:

1 Nominated subcontractors are employed.
2 The engineer is a member of the employer's staff.

In the first case, the employer, through his agent (the engineer), instructs the main contractor to place contracts based on quotations received, with specified subcontractors. Those subcontractors will then have contracts with the main contractor – not the employer.

In the second case, the employer and the engineer are effectively one, although it should be understood that they are contractually independent, and that in matters of dispute between the employer and the contractor the engineer is required to be strictly fair and to act in an impartial manner. This situation arises particularly where work is carried out for central government departments, local authorities, statutory undertakers and nationalized bodies.

Discharge of contract

Contracts are said to be discharged when the contracting parties are released from their contractual obligations.

Discharge does not come about automatically, but is usually brought about by an act of one or both of the parties. Once the contract is discharged the parties are no longer bound by its terms, although the discharge itself may result in enforceable rights. It follows that the precise time of discharge is important. Situations may arise in construction contracts where the contractor's work is terminated but the contract is not discharged. An example of this is covered by the determination clause (see Chapter 10). Likewise, if hidden defects, attributable to the contractor, are found in construction after completion of the contract works but during the period of limitation (see page 5), the contract will not be discharged and the contractor will be held liable.

There are four ways in which a contract may be discharged:

1 *Performance* where the parties have performed all their respective obligations. For example, the contractor has carried out the work fully in accordance with the contract, and the employer has paid all amounts due to the contractor.
2 *Frustration* where the parties are unable to perform their obligations owing to events outside their control. For example, Government action holding up the work for a lengthy period, hence radically changing the nature of the resumed contract. It should be noted that contracts are not frustrated if they become too difficult or too expensive to perform.

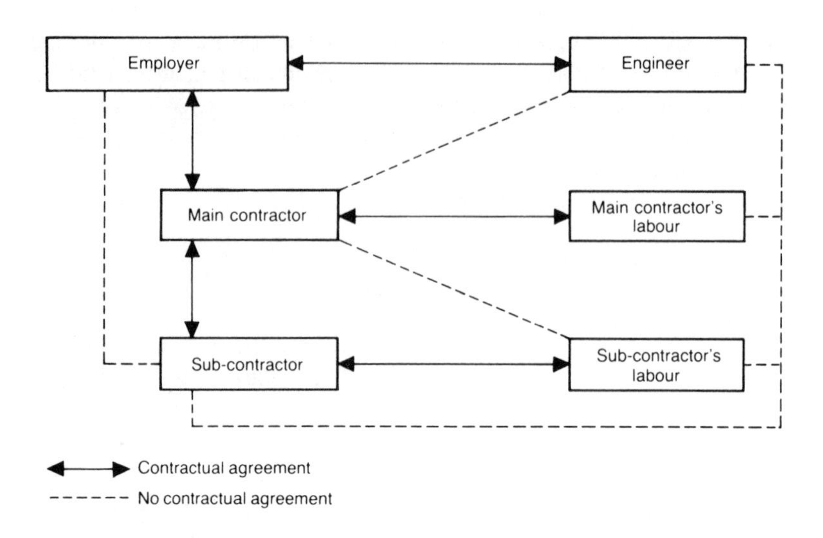

Figure 1.4 *Contractual relationships*

3 *Breach* where one party fails to perform an obligation (see page 11).
4 *Agreement* where both parties agree to terminate the contract before complete performance owing, perhaps, to changed circumstances. For example, excessive variations which have changed the nature and scope of the contract.

2 Standard forms, types of contract and contract documents

Reference ICE Clauses: 5, 6, 7 and 10

No two civil engineering contracts are identical. Indeed, probably no two construction contracts, either civil engineering or building, are truly the same. Identical conditions of contract are not likely, therefore, to be required. However, for work of a similar type, certain conditions will apply in the vast majority of cases. In fact, the same conditions, with only a few exceptions, are likely to be required for all but a few contracts. It follows, therefore, that a standard form of conditions for a given type of work, for example civil engineering, will remove the necessity of thinking out and drafting new sets of conditions for every new contract. By taking that standard form and modifying it to suit the requirements of a particular contract, time and effort will be saved while retaining the qualities of conditions which would be prepared otherwise. Amendments to the standard form ought then to be kept to the minimum necessary to meet the intentions of the contracting parties, and should be carefully drafted so as not to deviate too far from the parent document, or to introduce ambiguities and contradictions.

Standard forms of conditions are prepared jointly by professional bodies and organizations representing contractors, or by large organizations and public bodies to suit their own circumstances. The intention is that a common approach by the parties to all contracts will be achieved and, likewise, a standard interpretation of the risks and responsibilities involved.

Principal standard forms of contract used in civil engineering

There are a number of standard forms of conditions of contract used in civil engineering. Those most commonly used are:

ICE Conditions of Contract (sixth edition, 1991)
This document includes the *Forms of Tender, Agreement and Bond for use in connection with Works of Civil Engineering Construction*, and is issued jointly by the Institution of Civil Engineers, the Association of Consulting Engineers and the Federation of Civil Engineering Contractors. It is applicable to all civil

engineering construction works, and is the standard form to which later chapters of this book specifically refer. It is particularly suitable for general civil engineering work which is predominantly either in the ground or in, or adjacent to, water, and caters for the attendant risks and claim situations. It is also used, sometimes, for building works, and for mechanical and electrical works where such works are included in a civil engineering or building contract.

ICE Conditions of Contract for Minor Works (1988)

This document was sponsored and approved jointly by the Institution of Civil Engineers, the Association of Consulting Engineers and the Federation of Civil Engineering Contractors, and is restricted to only twelve clauses.

It is intended for use on contracts where:

1 Potential risk to the contracting parties is reckoned to be small.
2 Contract duration does not exceed six months, unless the basis of payment is daywork or cost plus fee (see below).
3 Works are simple and straightforward.
4 The contractor has no design responsibility except, possibly, for specialist work.
5 The contract does not exceed £100,000.*
6 The design is complete before tenders are invited.
7 Nominated subcontractors are not employed.

This is a versatile form of contract providing payment to the contractor by any of the following methods:

1 Lump sum.
2 Measure and value, using a priced bill of quantities.
3 Schedule of rates (with approximate quantities of the major items).
4 Daywork.
5 Cost plus fee.

Two or more of the above methods of payment may be used on one contract.

The following points should be noted:

1 Contracts should normally be fixed price, i.e. without provision for price fluctuations.
2 The Conditions of Contract should not be amended or extended in any way.
3 Acceptance should be within two months of the submission of tenders.
4 A named individual should be appointed as Engineer who, according to the Notes for Guidance, 'will personally be responsible for the Works'. He should not normally delegate his powers.

*The contract is suitable for larger amounts providing the scheme is straightforward. Due allowance should be made for inflation.

5 Liquidated damages should be limited to 10% of the estimated final value of the contract.
6 Daywork should be paid in accordance with rates current at the date the work is carried out.
7 Retention should normally be 5% with a limit of between 2½–5% of the estimated final contract value.
8 Minimum interim certificates should be 10% of the estimated final contract value rounded up to the nearest £1,000.
9 Minimum third party insurance cover of £500,000 for any one accident/unlimited number of accidents should normally be insisted upon.
10 Defects correction period should normally be six months. (This differs considerably from the twelve months maintenance period of the ICE fifth edition, and the unspecified defects correction period of the sixth edition.)
11 Disputes may be settled by conciliation, using the ICE Conciliation Procedure 1988. The object is settlement with the minimum delay, but relies on both parties wishing to reach agreement. Alternatively, the parties may refer the dispute to arbitration conducted in accordance with the ICE Arbitration Procedure 1983 (probably Part F, the short procedure).
12 Termination of the contract or determination of the contractor's employment is not provided for.

Although the contractor is required to proceed 'with due expedition', there is no facility to sanction a contractor who:

(a) is very slow or dilatory
(b) abandons the scheme
(c) becomes bankrupt
(d) goes into liquidation

or, indeed, an engineer who fails to certify, or an employer who fails to pay the contractor.

Federation Internationale des Ingenieurs-Conseils (FIDIC)
These conditions take two forms and are, in effect, international versions of the ICE and the IMech.E/IEE conditions (see page 20) to which they are closely related.

The fourth edition (1987) of the Conditions of Contract for Works of Civil Engineering Construction comprises:

Part I. General Conditions with Forms of Tender and Agreement.
Part II. Conditions of Particular Application with Guidelines for Preparation of Part II clauses.

The document is printed in several languages, although the English version is considered by FIDIC as the official and authentic text for the purpose of translation.

It is intended for general use for works where tenders are invited on an international basis, but is also suitable for use on domestic contracts.

The objective is to provide a standardized document which is both well known, internationally recognized and accepted, and adequately reflects the interests of the parties concerned.

Part I contains clauses of general application to suit any project, and is generally similar in content and order to the ICE fifth and sixth editions. Additionally, the order of priority of the contract documents is stated.

Part II is linked to Part I by corresponding numbering, hence the two Parts together comprise a single document governing the rights and obligations of the parties. Part II must be specially drafted to suit each individual contract, and its clauses may arise for one or more of the following reasons:

1 The wording in Part I specifically requires further information to complete the Conditions.
2 The wording in Part I indicates that supplementary information may be included in Part II, although the Conditions would still be complete without such information.
3 The type, circumstances or locality of the Works necessitates additional clauses or subclauses.
4 The law of a country or exceptional circumstances necessitate an alteration in Part I.

General Conditions of Government Contracts for Building and Civil Engineering Contracts (GC/Works/1) – Edition 3

Unlike most other standard forms, this form is not produced by a joint body representing interested parties, but solely by the employing agency: the Department of the Environment.

The Standard Form of Contract – Lump Sum with Quantities* was published in December 1989 by Her Majesty's Stationery Office, and is used for major government contract works. It is also usable, with amendments, by other public or private sector clients who favour firm contract control.

This version is for use with bills of quantities where all or most of the quantities are firm, i.e. not subject to remeasurement, giving a lump sum contract subject to adjustment only for variations, instructions, prolongation and disruption, financial charges and provisional sums.

Key features include:

1 The requirement for a resource programme and for regular progress meetings.
2 Power to order acceleration of the Works.

*Additional separate versions are planned for the future resulting in two lump sum forms (bills of quantities, and management) and two measure and value forms (approximate quantities, and schedules of rates).

3 Procedures and rules for the valuation of major variations, including the option requiring the contractor to submit a lump sum quotation in advance.
4 Interim payments based on stage payment charts. Accepted and agreed variations, prolongation and disruption claims, and finance charges paid in full without deduction of retention.
5 Claims subject to strict time limits.
6 Cost saving incentives for contractors.
7 Bad weather at contractor's risk.
8 Generally, any decision of the authority (the employer) and its agents in the settlement of disputes is final and conclusive.

General Conditions of Government Contracts for Building and Civil Engineering Minor Works (GC/Works/2) – Edition 2, 1980

This form, similarly to GC/Works/1, was prepared solely by the employing agency and is published by Her Majesty's Stationery Office. It is intended for use specifically by Crown Departments on contracts up to £50,000 in value (although it is anticipated that this figure will be increased following the introduction of the third edition of GC/Works/1). It is not really suitable for other than central government contracts.

The form is relatively simple and clear with the contract based on a specification and drawings, and a fixed price which is not even adjusted to allow for increases in levies and taxes.* A bill of quantities is not required.

There are no recitals, articles or appendix, but the contract incorporates the tender and written acceptance together with an 'Abstract of Particulars' which identifies the employer and indicates the person designated as the Superintending Officer (SO).

The SO controls the works insofar as the contractor is required to carry out and complete the works to his satisfaction. Furthermore, his decisions (other than in the case of arbitration) are final and conclusive.

The final account is prepared by the contractor and sent to the employer as soon as possible after completion, and the final sum paid when the SO certifies that the works are in a satisfactory state following the maintenance period.

British Electrical and Allied Manufacturers Association (BEAMA)

There are a number of versions of these conditions of contract which are used by a federation of nineteen trade associations representing electrotechnical and allied manufacturers.

They are suitable when quoting for either the supply of equipment and plant,

*The contract sum may be adjusted nevertheless in accordance with the contract, i.e. on account of ordered variations and provisional sum work.

or the supply and supervision of erection of equipment and plant. The document wording is such that its use need not be restricted simply to electrical work.

Additional standard forms
In addition to the above mentioned standard forms used in civil engineering, other forms of a specialist nature exist such as the IMech.E/IEE *Model Form of General Conditions of Contract MF/1* for mechanical and electrical engineering works, and the *Model Form of Conditions of Contract for Process Plants* for chemical engineering.

The 1988 edition of the IMech.E/IEE conditions includes forms of tender, agreement, subcontract and performance bond, and is intended for home and overseas contracts.

It is recommended by the Institutions of Mechanical and Electrical Engineers and the Association of Consulting Engineers. It recognizes recent important changes in practice, particularly regarding the role of the Engineer, and incorporates special sections covering subcontracts, and electronics hardware and software. It is suitable, therefore, for lump sum contracts for both the supply and erection of most types of electrical, electronic and mechanical plant.

MF/1 is not suitable, unless amended, for contracts where substantial remeasurement is envisaged. In such cases Model Form E should be used.

Finally, mention should be made of the JCT Standard Form of Building Contract (1980 edition) – the building equivalent of the ICE Conditions. This is probably the most commonly used of all the standard forms and has six versions, three appertaining to local government contracts and three to private works. These forms are issued by the Joint Contracts Tribunal representing the Royal Institute of British Architects, the Building Employers' Confederation, the Royal Institution of Chartered Surveyors, the Association of County Councils and, in the case of the local authorities versions, seven other associations. The 1980 edition is intended for use with large building contracts. For works of a lesser or different nature the JCT has issued other standard forms. These include:

The JCT Intermediate Form of Building Contract IFC.84 (second edition 1989)
The JCT Agreement for Minor Building Works (1980 edition, revised 1989)
The JCT Standard Form of Building Contract with Contractor's Design (1981 edition)
The JCT Standard Form of Measured Term Contract (1989 edition)

Types of contract

There is a wide variety of types of contract used in civil engineering, and each type has specific characteristics with which it tends to be identified. Projects may be prepared under the heading of one type but could well include characteristics of more than just a single type. The following titles and descriptions refer to those most commonly found.

Fixed price contract

Building contracts under the JCT *Standard Form of Building Contract* are for an agreed sum – the 'contract sum' – and are varied only on account of ordered variations, revaluation of prime cost or provisional sums and claims. Effectively, such contracts are 'lump sum' contracts, regardless of whether or not they are based on bills of quantities, but care should be taken so as not to confuse this type with the genuine 'lump sum' contract described later in this chapter. Civil engineering contracts under the ICE conditions, on the other hand, are based on quantities which are subject to measurement as the work proceeds and are referred to as 'remeasurement' or 'measure and value' contracts, with the final figure in the tender being known as the 'tender total'. Such contracts are 'fixed price' contracts. If, however, the contract is subject to price fluctuations (see Chapter 7), it is not a fixed price contract, although the individual rates against item descriptions remain unchanged.

Bill of quantities contract

This is the most common, and generally the soundest, form of contract used for both civil engineering and building projects. It is used where the design has been completed and is based on the resulting drawings and specification, and incorporates a bill of quantities in which the project is broken down into individual items of work together with quantities for each item. Each item is subsequently priced by the contractor to arrive at the 'tender total'. The contract must indicate whether the quantities are firm or subject to remeasurement on completion of the site work. The measurement is standardized and regulated by the following documents:

1 For civil engineering works:
   (a) The *Civil Engineering Standard Method of Measurement* (second edition, 1985) published by the Institution of Civil Engineers.
   (b) The *Method of Measurement for Highway Works (1987)* published by HM Stationery Office.
2 For building works:
   (a) The *Standard Method of Measurement of Building Works* (seventh edition, 1988) published jointly by the Royal Institution of Chartered Surveyors and the Building Employers' Confederation.
   (b) The *Principles of Measurement (International) for Works of Construction (1979)* published by the Royal Institution of Chartered Surveyors.

This type of contract offers several advantages to the employer as follows:

1 A common basis for competitive tendering.
2 An estimate of the cost of the job.
3 A basis for payment of interim and final accounts.
4 A basis for valuing variations.

Additionally, contractors may use bills of quantities to:

1 Calculate quantities of materials for ordering, although this is a bad practice as the billed quantities are, first, net figures and, second, in the case of civils, only estimated (see Clause 55 of the *ICE Conditions of Contract*).
2 Obtain subcontractors' quotations.
3 Check that all the work carried out is covered by the contract.
4 Calculate bonus payments to site operatives.

Lump sum contract
In this type of contract the contractor undertakes to carry out the works for a fixed sum of money regardless of the component parts of the works. Drawings showing the details of the project are usually provided, together with a specification, to assist the contractor in pricing the job, but no bill of quantities is supplied. This type of contract is often used where the quantity and specification of the work are fully known and the works are small in extent. Examples include site clearance work and small constructional works. By agreement, a lump sum contract can be either fixed priced or subject to adjustment on account of price fluctuations.

Schedule contract
This type of contract is commonly used for maintenance repair and work, where quantities of work cannot be established with accuracy at the tender stage, and where the flow of work is likely to be irregular and, possibly, at various locations. As a result, this type of contract is often used by local authorities for 'annual tenders', whereby contractors undertake to do work for agreed rates for a specific period of twelve months. Schedules of rates are also used on occasions as a basis for negotiated contracts. Schedule contracts usually include approximate quantities to assist contractors in pricing and for the subsequent comparison of tenders, and take two forms:

1 The employer supplies a schedule of items of work together with unit rates against each item. Contractors then indicate a percentage variation, either up or down, that they require for carrying out the work.
2 The employer prepares a schedule as in 1 but excludes the unit rates, leaving contractors to insert their own rates against each item. This is the more usual method of the two.

Figure 2.1 illustrates a simple schedule of this type with approximate quantities and rates inserted by three contractors. The three sets of figures have been shown together for comparison. For simplicity, items have been given numbers rather than descriptions.

Negotiated contract
There is an element of negotiation in all contracts but a negotiated contract is one where the employer and a potential contractor negotiate a contract by way of

Item	Contractors								
	A			B			C		
	Rate	Quantity	Amount	Rate	Quantity	Amount	Rate	Quantity	Amount
1	£10	100	£1000	£13	100	£1300	£11	100	£1100
2	£ 5	160	£ 800	£ 6	160	£ 960	£ 7	160	£1120
3	£ 6	200	£1200	£ 5	200	£1000	£ 4	200	£ 800
4	£50	50	£2500	£45	50	£2250	£46	50	£2300
5	£14	200	£2800	£16	200	£3200	£17	200	£3400
Totals	£85		£8300	£85		£8710	£85		£8720

Figure 2.1 *Comparison of schedule contract tenders which include approximate quantities*

discussion rather than the contract being won by formal competitive tendering. Situations where this type of contract may be suitable are where the specification requirements are not fully known, where specialized plant or services are supplied by a single supplier, or where a nominated subcontractor is employed.

Care should be taken to prevent negotiations being protracted to such an extent that they preclude alternative arrangements being made. Areas of confidentiality should be established during negotiations so that in the event of an agreement not being finalized and the employer turning to another contractor, he is not guilty of making use of specialist information, etc. obtained from the original contractor. Time is not necessarily saved with this type of contract, although that may well be intended at the start of the negotiations.

Reimbursable cost contract

Contracts falling into this category are frequently referred to as 'cost-plus' contracts, whereby the contractor is reimbursed his actual costs plus his overheads and profit. No total contract price is quoted and so this type of contract is somewhat open ended, particularly if there is no constant and suitable supervision by a representative of the employer. Depending on the particular form that the contract takes, there is little or no incentive for the contractor to complete the works quickly or to try to reduce costs. On the other hand, this type of contract has the merit of being useful in an emergency when the time available to prepare a detailed scheme before work is commenced is insufficient, or when the scope and progress of the contract cannot be accurately foreseen and assessed. All the contractor's records and accounts must be made available for inspection by the employer's representative so that agreement on costs can be reached. There must also be agreement prior to the start of the work as to what constitutes 'costs', the rates of pay, what form the 'plus' will take, and who will bear the cost of 'making good' faulty work. Reimbursable cost contracts fall into three principal categories:

Cost plus percentage contract
The contractor is paid his actual costs as previously indicated plus an agreed percentage on the allowable costs. The percentage may be a single figure, or it may be various agreed figures relative to particular elements of cost such as labour, materials and plant. This is not a particularly good form of contract as it lends itself to abuse by contractors who may delay the completion of the works in order to keep their workmen employed or to increase their profits. This type of contract is frequently referred to as a 'daywork' contract.

Cost plus fixed fee contract
Again, the contractor is paid his actual costs, but this time he receives, in addition, a fee in the form of a fixed lump sum, instead of a percentage. This is not a particularly good form of contract, although it is an improvement on the percentage type, as there is some incentive to the contractor to complete the works quickly.

Cost plus fluctuating fee contract
Once again, the contractor is paid his actual costs, but the amount of the fee is determined by reference to a sliding scale which increases his fee relative to savings on the job. In other words, the lower the costs, the greater the fee. Thus, the contractor is encouraged to carry out the work as quickly and as cheaply as possible. From the employer's point of view, this type of contract is the best of the reimbursable cost contracts.

A variation on this type of contract is shown in Figure 2.2, whereby the contractor's fee is calculated on reducing percentages as the cost (prime cost) increases.

Target contract
The target contract is a development of the reimbursable cost contract which provides a positive incentive to the contractor to complete the work quickly and economically, but it involves a certain amount of document preparation including a bill of quantities.

Target contracts are used where tendering is required before detailed design has been completed and requires the total contractor's costs to be estimated as

Contractor's costs		Percentage addition	Contractor's fee
First	£25,000	15%	£3,750
Next	£25,000	10%	£2,500
Over	£50,000	5%	£ 750

The above assumes that the contractor's agreed costs total £65,000.

The contractor's fee would be £7,000 and the total cost to the employer £72,000.

Figure 2.2 *An example showing how a contractor's fee may be calculated in a cost plus fluctuating fee contract*

accurately as possible, even though the actual quantities may not be known at the commencement of the works. This figure may be ascertained by means of a brief bill of quantities drawn up as details become available and the total finally agreed after the work has commenced. The total estimated contractor's costs sets the target and fixes the fee.

The contractor is eventually paid his actual costs plus the fee. If the actual costs are less than the target cost, he receives an addition to the fee based on an agreed proportion of the difference, or on a sliding scale. Conversely, if his costs exceed the target cost he receives only an agreed proportion of the excess cost which, in effect, means a reduced fee. The target cost may be adjusted to allow for variations in design and quantity, and for fluctuations in labour and material costs.

Various versions of target contracts exist whereby the calculation of the fee finally paid may vary from the foregoing. One example is where the fee remains unchanged, other than on account of variations and the like, but any cost in excess of the target cost is borne by the contractor. If, on the other hand, the cost is less than the target, the employer receives the entire benefit.

Serial contract and serial tender

A serial contract is a contract whereby the parties are firmly committed to a series of specified projects following an initial contract.

A serial tender is similar to the serial contract, but takes the form of a standing offer from a contractor to enter a series of separate contracts spread over a specified period. It may be used for the supply of certain materials, or for the performance of work to a stated minimum value. It enables the employer to call on the contractor for goods or services without having to negotiate a new contract each time one is required, does away with the need to have a firm total requirement known in advance, and often leads to a more favourable price than could be expected if each of the contracts was negotiated individually.

All-in contract

This type of contract is also known as a 'package deal' contract, because it combines two or more related tasks, each of which could form a contract on its own. It may combine design with construction, development with construction, or design and development with construction. Likewise, design and development may be combined with supply and erection or maintenance.

The usual procedure with this type of contract is that the employer outlines his requirements to the contractor, who then designs and constructs the project for a given sum of money. Civil engineering contracts by their very nature do not lend themselves readily to this type of contract, although all-in contracts may be used for special types of work such as power stations and petro-chemical works. If both design and development are included in a particular contract, the method of payment is sometimes on a 'cost-plus' basis, or based on schedules of rates.

The following points are generally considered to be the advantages of the all-in contract:

1 Reduced design and tendering time.
2 Reduced overall costs.
3 The contractor's specialist knowledge and experience can be fully utilized.
4 The contractor is likely to get a better price for carrying out the works which may result in better workmanship, etc.
5 There is continuity of responsibility, reducing the likelihood of claims.
6 Contractors may undertake unattractive projects that are linked with favourable ones.

Disadvantages associated with this type of contract are:

1 Comparison of tenders is difficult.
2 The savings could be lost if the work is not properly controlled.
3 There is no one to protect the employer's interests, unless special arrangements are made.
4 The contractor may choose the most favourable design for him – not necessarily the best for the employer.
5 The contractor may use subcontractors who would not be 'approved' for other contracts.
6 The employer has no choice of subcontractors, as he has with nominated subcontractors in other contracts.

Management contract

Management contracts are the result of a relatively recent fresh approach to the placing of construction contracts. The traditional manner of carrying out large projects, with an independent engineer, a properly detailed design, and a contractor chosen by competitive tender often fails to provide the best working situation and relationship for the engineer and the contractor. Sometimes, it also results in the employer not getting the best, or even fair, value for his money; it being argued that the separation of the design and construction functions makes it difficult to control the work once the contract has been let. Management contracts are used as a means of reducing these difficulties.

In a management contract the contractor joins the employer's/engineer's team at the design stage of a contract to assist in the planning and design by providing the team with the benefit of his practical experience and construction expertise. Once the project gets under way, he manages the contract but does not carry out any of the construction work, leaving that to be carried out by various subcontractors who are sometimes referred to as direct works contractors. The subcontractors are chosen by the design team following submission of competitive tenders from a select list of specialist subcontractors. The management contractor is paid a fee based on a fixed percentage of an agreed estimated value of the works for managing the contract, plus his costs incurred during the design stage. The role of the engineer remains basically unchanged, being fully responsible for the design, issue of variations and agreement of accounts, etc.

One of the principal advantages claimed for this type of contract is that the project time is reduced, because subcontractors' quotations can be obtained immediately after completion of the relevant design, rather than having to wait for the complete project design as in a conventional contract. Once such quotations have been obtained, the work can commence providing all necessary preliminary works have been completed. A second benefit is that a joint engineer/contractor team effort creates a unity of purpose, hence reducing the likelihood of 'claims', and improving the prospects of a better end product. However, claims may still be forthcoming from subcontractors, and the employer does not have the benefit of knowing the overall contract cost and duration prior to commencement of the works.

Construction management contract

This type of contract operates in a similar manner to the management contract, but differs in that the employer has contracts directly with the various subcontractors leaving the contractor to manage the works as the agent of the employer.

Primary clauses

A standard form of conditions of contract, as previously mentioned, will include conditions that are common to all contracts of a given type. Indeed, it can be safely assumed that all construction contracts have the same basic requirements, although they may differ in detail and total content. For instance, employers will be concerned with that which they expect to get out of the contract, together with the quality of work and materials and the date of completion. Contractors, for their part, will be concerned with payments and profit, particularly the latter. After all, profit-making is their reason for being in business: the construction work being simply a means to that end. They will also want to know if retention is to be held, and if so, how much and for how long. Furthermore, what happens to payments due in the event of the contract overrunning, and for which contributory factors will they be held responsible?

These basic requirements result, then, in clauses appearing in contracts which in general are similar, although their details may vary in different standard forms and, maybe, in different contracts. Because such clauses are an essential requirement of a contract they are often referred to as 'primary clauses'.

Primary clauses are dealt with in detail later in this chapter, or elsewhere in following chapters, and can be classified under the following headings.

Instructions regarding work

These include general obligations, workmanship and materials (*ICE Conditions of Contract*, Clauses 8–40 and 58(2)).

Additional instructions that need to be drawn to the contractor's attention when pricing the bill of quantities are included in the invitation to tender under the heading of 'Instructions to tenderers' or 'Instructions for tendering'.

Contract documents

The documents that are recognized as contract documents are stated in the form of agreement. The relevant ICE clauses are Clauses 5–7.

Variations to the contract

The procedure for ordering variations on account of alterations, additions and omissions, and their subsequent valuation is covered by ICE Clauses 51 and 52.

Possession of site, completion of work and extensions of time

In addition to possession, completion and extensions of time, under this heading are likely to be included the requirements in relation to commencement of the works, liquidated damages and the certificate of substantial completion, all of which are dealt with by ICE Clauses 41–48.

Measurement and valuation

The clauses cover quantities, measurement requirements, valuation and the method of measurement adopted. The relevant ICE clauses are Clauses 55–57.

Certificates and payment

Procedures with regard to interim and final accounts submission are set out here, together with the requirements in respect of retention, and the defects correction certificate. These are covered by ICE Clauses 60 and 61.

Contract documents

The majority of civil engineering contracts tend to be of the bill of quantities type, and are compiled to a standard format generally as follows:

1 Form of tender
2 Appendix (to the form of tender)
3 Form of agreement
4 Conditions of contract
5 Special conditions
6 Specification (including a list of contract drawings)
7 Preambles to the bill of quantities (and measurement clauses)
8 Bill of quantities

Examples of 1, 2 and 3 can be found at the back of the *ICE Conditions of Contract* (pages 47–51), and it is in one of these, the form of agreement, that the documents deemed to be contract documents are listed. They are:

1 The tender and written acceptance
2 The drawings
3 The conditions of contract
4 The specification
5 The priced bill of quantities

The recognized order of importance of the contract documents in civil engineering is conditions of contract, drawings, specification, bill of quantities.

The form of agreement records the parties to the agreement, and the date on which the agreement was made. It identifies the work to be carried out, and states that the employer has accepted the contractor's tender to construct and complete the works. The contract documents are identified, both parties agree to abide by the conditions of contract, and the employer undertakes to pay the contractor for the work carried out.

The form of tender and appendix

The tender is addressed to the employer, and is the contractor's formal offer to construct and complete the permanent works in conformity with the drawings, conditions of contract, specification and bill of quantities, within the time stated in the appendix. It includes an undertaking that, if the tender is accepted, the contractor will provide performance security (enter into a bond). The security must be provided by a body approved by the employer and be for a sum not exceeding 10% of the tender total. (See Clause 10.)

The appendix is in two parts:

Part 1 to be completed prior to the invitation of tenders
Part 2 to be completed by the contractor

Part 1 includes:

1 Name and address of the employer
2 Name and address of the engineer
3 Defects correction period
4 Number and type of copies of drawings to be provided
5 Contract agreement
6 Performance bond
7 Minimum amount of third party insurance
8 Works commencement date
9 Time for completion
10 Liquidated damages for delay
11 Vesting of materials not on site (employer's option)
12 Method of measurement adopted
13 Percentage of the value of goods and materials to be included in interim certificates
14 Minimum amount of interim certificates
15 Rate of retention
16 Limit of retention
17 Bank whose base lending rate is to be used
18 Requirement for prior approval by the employer before the engineer can act

Part 2 includes:

1 Insurance excesses
2 Time for completion (if not completed in Part 1)
3 Vesting of materials not on site (contractor's option)
4 Percentage(s) for adjustment of PC sums

The drawings

The contract drawings show the nature and scope of the required project work and, ideally, should be prepared completely before the contract goes to tender. Where this is not possible, the general arrangements and as many of the details as will permit the contractor to price the bill of quantities should be produced, and forwarded to the contractor at the tender stage.

Drawings should contain sufficient descriptive and explanatory notes and dimensions to ensure that the quantities are taken off accurately and that work will be set out correctly. Particular care should be taken that new and existing work is clearly identified, that the nature of the site and ground conditions are indicated and that, where applicable, water tables, existing sewers, services and foundations, etc. are shown.

Once the contract work is under way, further drawings, particularly details, are likely to be required. These should be produced and sent to the contractor without delay in order that he may complete the work speedily and satisfactorily.

The conditions of contract

The conditions of contract have been described under 'Principal standard forms of contract', and will be referred to again in later chapters.

It should be pointed out, however, that following the 'Form of Tender', contracts usually include 'Conditions of Contract' under which modifications and additions to the standard form are included. These precede 'Special Conditions', which contain any special requirements of the contract, particularly in relation to statutory bodies and the like.

The specification

The specification informs the contractor of the precise requirements of the engineer in respect of the various operations and areas of work contained in the contract. The drawings show the locality, layout, dimensions and brief descriptions of the work but, generally, little more, for there are obvious limitations to the amount of detailed information that can be included on any drawing. Clearly, if all the engineer's requirements were included, there would soon be more written information than drawing, and so these requirements are set out in a separate document: the specification.

The drawings show what is to be constructed, whereas the specification tells the contractor how it is to be constructed, the standards required, and the quality of workmanship and materials, etc.

To aid the engineer preparing a contract there are standard specifications. Additionally, many organizations have their own libraries of specification clauses, covering the areas of work within their particular field. These have usually been built up over a period of years.

Other organizations that deal with particular types of work have their own specifications: British Rail, for instance, and the Department of Transport (DTp.). The DTp. Specification for Highway Works, and the Highway Construction Details are in book form and are incorporated in most motorway contracts and local authority contracts for road and bridge works. With them there is a companion book, *Notes for Guidance*, which is to assist engineers preparing contracts. It explains how to use the specification and the reasons where necessary, for certain items and notes on individual clauses. It also includes an appendix of information required to be included when preparation is in accordance with the recommendations of the DTp.

Some specification clauses may not be included in the standard specification or may not be strictly applicable. Where this is the case it is necessary for the engineer to write additional or substitute clauses. The standard specification usually includes a preamble to the effect that 'Any clauses in this specification which relate to work or materials not required by the Works shall be deemed not to apply'.

Besides libraries of specification clauses and particular specifications of organizations or bodies there are other sources from which clauses are drawn or compiled. These include:

1 Previous contracts
2 British Standards
3 Codes of Practice
4 'Specification'
5 Manufacturers' and suppliers' technical literature

Clearly, the specification is an extremely important document. Any additional or substitute clauses must be written with care and only after considerable thought to their ultimate interpretation, for once written the contractor is deemed to have priced the bill of quantities in accordance with it. If a clause is wrongly worded it could result in the employer getting (and paying for) considerably more than required or, conversely, getting considerably less. In the latter case the resident engineer is likely to issue a variation order amending the specification, the contractor is likely to ask for additional money and the employer finish up paying more for the item than he would have done originally. Invariably, incorrect specification writing increases the final cost of the contract.

The foregoing, where the contractor is told the methods and manner required to perform the construction task, is known as 'Method Specification'.

An alternative type of specification is the 'End Result Specification' where only the required end product and its capabilities form the specification. The contractor is left, in this case, to construct in his own way and to use materials of his choice and the mode and manner of working that is most expedient to him.

The bill of quantities

A bill of quantities is a schedule of items of work required under the contract, with quantities against the majority of items. The quantities are normally measured in accordance with one of the standard methods of measurement mentioned previously, and should be as accurate as possible, although it should be understood that they are estimated and must not be taken as accurate (see *ICE Conditions of Contract*, Clause 55(1)). Descriptions must indicate clearly the nature and scope of the work covered, and are compiled as set forth in the method of measurement adopted for a particular contract. For instance, under the Civil Engineering Standard Method of Measurement each description is made up by taking one feature from each of three divisions. Alternatively, descriptions may be kept brief with reference being made to a particular drawing detail or, as in the case of, say a headwall, a type number.

The unit rates entered by contractors against items in civil engineering bills of quantities normally include *inter alia* for overheads, profit and, where required, for items of temporary works, such as timbering, keeping excavations free from water, etc. However, special items of exceptional temporary work, such as sheet piling or the diversion of a river, may be included at the discretion of the person preparing the bill of quantities.

Figure 2.3 illustrates a page from a bill of quantities prepared in accordance with the CESMM, the format of which differs slightly from the traditional layout in that the 'unit' and 'quantity' columns are reversed. The numbers in the first column indicate how the description is made up: the letter corresponds to the work classification, and the three digits give the positions of the features in their respective divisions.

The objects of a bill of quantities are stated in the CESMM as:

(a) to provide such information of the quantities of work as to enable tenders to be prepared efficiently and accurately.
(b) when a contract has been entered into, to provide for use of the priced Bill of Quantities in the valuation of work executed.

The arrangement of a civil engineering bill of quantities will vary with different projects and the layout of the works, in compliance with the CESMM. There it states that in order to obtain the above objects:

> Work should be itemised in sufficient detail for it to be possible to distinguish between different classes of work, and between work of the same nature carried out in different locations or in any other circumstance which may give rise to different considerations of cost.

The bill of quantities tends to be split, therefore, into sections for each component part of the works. Within each section, the work normally follows the order of the work classifications (or sections) contained in the method of measurement, and the individual items follow the sequence of the particular work classification into which they fall.

Finally, reference is made to *operational bills*, and to *method-related charges* included in bills of quantities measured in accordance with the CESMM. These have resulted from new techniques and methods in civil engineering construction and management. They reflect the variables in the cost of construction; and method-related charges, as now used in civil engineering bills, provide opportunities for tenderers to enter in the bill items that are directly related to their methods of construction, the costs of which are considered as not proportional to the quantities of other items and for which other rates and prices have not allowed. These items may be for accommodation and buildings, services, plant, temporary works and for supervision and labour. Under the CESMM, the tenderer must distinguish between those charges that are 'time-related' and those that are 'fixed'. Such items must be fully described, precisely define the extent of the work covered, identify the resources to be used and the particular items of work to which they relate. The contractor is not obliged to adopt the methods stated in items, and method-related charges are not subject to admeasurement. Therefore, amounts entered by the contractor are paid in full subject only to

CLASS E : EARTHWORKS

ROADS

Number	Item description	Unit	Quantity	Rate	Amount £	p
	EARTHWORKS					
E210	Excavation of cuttings topsoil for re-use	m^3	2,573			
E240.1	Excavation of cuttings material for disposal other than topsoil, rock or artificial hard material; Excavated surface 0.25m above final surface	m^3	96,317			
E240.2	Excavation of cuttings material for disposal other than topsoil, rock or artificial hard material; Commencing surface 0.25m above final surface	m^3	125			
E511	Excavation ancillaries trimming of slopes natural material other than rock	m^2	11,305			
E521	Excavation ancillaries preparation of surfaces natural material other than rock.	m^2	14,073			
E631	Filling and compaction thickness 150mm excavated topsoil	m^2	17,155			
E632	Filling and compaction thickness 150mm imported topsoil	m^2	2,635			
E831	Landscaping grass seeding to surfaces inclined at an angle not exceeding 10° to the horizontal.	m^2	8,485			
E832	Landscaping grass seeding to surfaces inclined at an angle exceeding 10° to the horizontal	m^2	11,305			
			Page total			

Figure 2.3 *Page from a bill of quantities prepared in accordance with the* Civil Engineering Standard Method of Measurement

changes arising from variations ordered by the engineer.

A list of suggested method-related items is shown in Figure 2.4.

The use of method-related charges for general obligations results in definite advantages compared with the traditional use of 'Preliminary Bills' (or 'Sections').

These can be summarized as follows:

1 More equitable payments in interim certificates, leading to improvements in the contractor's 'cash-flow' position.
2 Easier evaluation of costs arising from changes in methods of construction and timing brought about by variations.
3 Reduced areas of dispute when quantifying claims.
4 Less need to vary rates under Clauses 52 and 56(2) (see Chapter 7).

ICE requirements

Clause 5 requires all the contract documents to be mutually explanatory, and where ambiguities and discrepancies occur they must be explained and adjusted by the engineer, who must then instruct the contractor accordingly in writing.

Under Clause 6, the contractor upon being awarded the contract should be supplied, free of charge, with four copies of the:

1 conditions of contract
2 specification
3 unpriced bill of quantities
4 number and type of copies of all the drawings entered in the appendix and listed in the specification

Further copies may be obtained by the contractor at his own expense.

Similarly, where the contractor is required to design part of the permanent works he must, upon approval of the design, supply the engineer with four copies of the:

1 drawings
2 specifications
3 other documents including:
   (a) the necessary calculations and other information previously submitted to the engineer to satisfy him as to the suitability and adequacy of the design.
   (b) operational and maintenance manuals and drawings, where applicable.

Further copies may be obtained by the engineer at the employer's expense upon written request.

All the said drawings, specifications, unpriced bills of quantities and documents remain the copyright of the supplier. As a consequence, all copies of the engineer's drawings etc. should be returned to him at the end of the contract, but

ITEM NUMBER	ITEM	TIME-RELATED	FIXED
	Accommodation and Buildings		
A.311	Offices	√	√
A.314	Stores	√	√
A.315	Canteens and messrooms	√	√
	Services		
A.321	Electricity	√	√
A.322	Water	√	√
A.323	Security	√	
A.325	Site transport	√	√
A.326.1	Personnel transport - vans	√	
A.326.2	Personnel transport - cars	√	
A.329	Telephone	√	√
	Plant		
A.331	Cranes	√	√
A.332	Transport		√
A.333	Earthmoving	√	√
A.335	Concrete mixing	√	√
A.336	Concrete transport	√	√
A.339	Sundry small plant, eg saw benches, welding gear etc.	√	√
	Temporary Works		
A.351	Traffic diversions		√
A.352	Traffic regulation	√	
A.353	Access roads		√
A.354.1	Bridges - bailey	√	√
A.355	Cofferdam		√
A.356	Pumping	√	√
A.357	De-watering	√	√
A.363	Piling	√	√
A.366	Hardstandings		√
	Supervision and Labour		
A.371	Supervision	√	
A.372	Administration	√	
A.373.1	Labour teams - unloading gang	√	

Figure 2.4 *Method-related charges – a list of the most commonly used items*

the employer and the engineer are entitled to reproduce and use the contractor's design details etc. for the purpose of completing, operating, maintaining and adjusting the works.

The engineer, in accordance with Clause 7, must supply the contractor as required during the progress of the works with any modified or additional drawings, specifications and instructions that the engineer considers are necessary for the proper and adequate construction and completion of the works, and the contractor is obliged to carry out the requirements of such drawings etc. If the foregoing require any variation to the works, the variation will be deemed to have been issued.

Where the contractor has designed part of the permanent works, he must also supply such further documents as the engineer considers necessary and subsequently will be bound by the engineer's approval of them.

Should the contractor require further drawings etc. in order to carry out the works, he must give the engineer notice in writing, and if the engineer fails to issue the required drawings etc. within a reasonable time (and providing the engineer considers that they are really necessary), the contractor should be reimbursed for any reasonable additional costs incurred, and the contract extended to allow for the delay.

If the failure is due in some way to the contractor, the engineer should take such into account when determining the amount of reimbursement of the additional costs incurred and the extended time for delay.

A copy of the drawings and specification issued to the contractor, and a copy of all the contractor's design drawings, specifications and documents must be kept on site and should be available at all reasonable times for the inspection and use of the engineer and any persons authorized by him in writing.

The engineer is responsible for the integration and co-ordination of the contractor's approved design with the rest of the works, but the approval of the contractor's design does not relieve the contractor of his contractual responsibilities.

3 Selection of main and subcontractors – tendering procedures

Reference ICE Clauses: 1, 4, 11 and 59

Generally, the selection of contractors for civil engineering contracts follows a fairly lengthy, established procedure, but, if the circumstances of a particular project so warrant, a simpler and more expeditious form of selection may be used. Whether or not a short cut method is adopted, however, may well depend on the type of employer for, if the employer is an authority from the public sector, the difficulties of justifying such an approach may be too formidable, bearing in mind the need to comply with standing orders and for public accountability.

A prerequisite to choosing a contractor is the need to obtain a tender, which can be defined as an offer by a tenderer to enter into a contract for a price on stated terms, following an invitation from a prospective employer, or engineer on his behalf.

Tenders for the supply of goods and services fall into two classes, those where the tender is:

1 For the supply of a specified or definite quantity.
2 A standing offer to supply goods or services from time to time as required.

Every tender for the supply of goods or services is a separate offer, and each one may or may not be accepted. Once accepted a contract is formed (see Chapter 1).

If there is a standing offer for the supply of goods and services as required, every time an order is placed a separate acceptance is made.

Generally, if the purchaser does not give an order or, alternatively, does not place orders for the full quantity of goods set out in the tender, a breach of contract has not occurred. Where, however, the buyer undertakes to buy all the goods he needs from the tenderer, he will break the contract if he obtains some of the goods from another source.

Tender lists, selective tendering, open tendering and negotiated tendering

Contractors are awarded contracts, primarily, by one of two methods of tendering: selective tendering, or open tendering.

Selective tendering

Under this method an invitation is sent to a list of suitable contractors, inviting each of them to tender for a particular contract. The list may be drawn up from either:

1. An *ad hoc* list of contractors of established skill, integrity and proven competence. The list will also depend on various other factors, namely, the nature of the project, its location, size and approximate value, and its special requirements, together with the availability of contractors to take on new work. As a result, the number of tenderers on the list may be reduced to a less than ideal figure.
2. The employer's/engineer's standing list of approved contractors. From this list the required number of tenderers is drawn for each new contract. The tenderers may be especially selected for the particular contract (maybe by computer), chosen at random or in strict rotation. The chosen method depends on the engineer and the policy of his organization.

It is generally recognized that the list of tenderers should not exceed six. If it is too long, there will be wasted time and expense in providing sets of enquiry documents, each of which could be extremely substantial. Furthermore, since only one tenderer can be successful, there will be an unnecessary amount of abortive pricing on the part of the unsuccessful contractors, which will be reflected eventually in their rates for future work.

If, on the other hand, the list is too short it might result in insufficient competition to arrive at a genuine commercial price for the job, and could lead to collusive tendering on the part of unscrupulous contractors.

Two possible disadvantages with selective tendering are, firstly, that it tends to result in more costly contracts than open tendering and, secondly, that contractors not on approved lists sometimes find difficulty in becoming approved.

It is recommended that preliminary enquiries are made to establish those contractors wishing to tender for the contract, and those who do not wish to be considered on this occasion. The enquiry should make it clear that a rejection of the invitation will not debar contractors from future work, and that some contractors who have notified their desire to tender, might find that they are not included on the final list of tenderers. Where this situation occurs, such tenderers should be informed as soon as possible after the list has been drawn up.

Open tendering

Under this method, contractors are invited by means of notices placed in suitable publications, such as technical and trade journals, and the national and local press, to apply for the tender documents and an opportunity to price a particular contract. Such notices give certain basic information about the project, for example its character, location, size, approximate value, the required completion date, and any other relevant points of particular interest to would be tenderers.

Applicants are likely to be asked to pay a non-refundable deposit for the contract particulars, partially to defray the cost of producing the documents, and to show that they are seriously interested in tendering for the work. Following receipt of the documents, the tenderer will price the bill of quantities and, generally, the employer will accept the lowest tender, provided that he is satisfied that the contractor has the necessary resources available to do the work, is reputable, and has successfully completed work of a similar nature. In many instances the employer will require the contractor to provide a performance bond.

Open tendering is considered to be a way of offering work fairly and avoids suggestions of favouritism towards certain contractors, but it suffers the following disadvantages when compared with selective tendering:

1 Advertising is necessary and more sets of contract documents are likely to be required, hence the tendering process is more costly.
2 The tendering period is protracted, which not only delays commencement and completion of the works, but may also increase the cost.
3 There is a risk that the contract will be awarded to an unscrupulous contractor.
4 There is a possibility of unsatisfactory performances by contractors not previously used, or that have limited or doubtful experience and resources.
5 In some instances, contractors may have to make assumptions to cover extra unknown factors when tendering, resulting in the employer not getting a realistic price.

The employers most likely to adopt this method of tendering are local authorities, nationalized industries and similar bodies, with private companies and organizations being more inclined towards selective tendering. Open tendering is frequently used, however, for the supply of certain bulk commodities required on a frequent and regular supply basis.

Since the United Kingdom became a member of the European Economic Community (now known as the European Community), proposed public works, and supply contracts for certain specified authorities have to be advertised in the *Official Journal of the European Communities* if they are estimated to cost in excess of a set monetary value. That value is reviewed from time to time, and the national currency equivalents recalculated every two years. Currently (1991), that value for works contracts is 5 million European Currency Units equivalent to £3,310,000. Thus, an opportunity is provided for contractors in all member states of the Community to tender for the work. The specified authorities include county and district councils in England and Wales, the Council for the Isles of

Scilly, the London Borough Councils and the City of London, the New Town Development Corporations, and the Commission for the New Towns.

Negotiated tendering
This is an alternative to selective and open tendering that is used relatively infrequently. Here a single preferred company makes a tender offer when the specification and contract requirements are known. The offer is usually preceded and followed by discussions, and is made subject to negotiation, as the name suggests. Items that are likely to be negotiated include the ultimate price for the works or the means of arriving at the price, the form of contract to be used, the programme and method of carrying out the works.

Tendering by this method is adopted in special circumstances where, for example, time restrictions prevent other forms of tendering, or the preferred company has performed successfully for the employer on previous occasions, or specializes in the type of work to be undertaken, or has special knowledge and experience of the site, its locality and conditions. (See also page 22.)

Single stage tendering: enquiry, checking and appraisal, negotiations, letter of intent and letter of acceptance

Single stage tendering is the method most commonly used to obtain tenders for civil engineering contracts, and is employed for both selective and open tendering. The procedure generally adopted is as follows.

Once the contract documents have been completely prepared and the list of tenderers drawn up, copies of the documents are sent to each of the selected tenderers. Accompanying the invitation to tender will be instructions to tenderers probably entitled 'Instructions for Tendering'. The 'Instructions' contain directions and advice for persons responsible for producing the tenders. The number and extent of such instructions will vary depending on the nature and complexity of the contract, but certain items appear in the majority of cases. They include:

1 A directive that details of the tender documents are to be treated as private and confidential.
2 A statement that the contract includes, or does not include, the contract price fluctuations clause. If it does, the particular one will be indicated.
3 An instruction that the tender be made on the form of tender, that it must be signed and submitted with the fully priced bill of quantities which must be totalled in ink, and returned with the other documents to a specified address by a specific date and time.
4 A warning that alterations or additions to any of the contract documents, or the bill of quantities not being properly completed, could result in the tender being rejected.
5 A directive that tenderers must complete and return a certificate certifying that they have not produced collusive tenders.

6 A reminder that insurers should be informed of any special insurance requirements under the contract.
7 A notification drawing the tenderer's attention to the fencing requirements and the need to indemnify the employer against damage to persons and property, etc.
8 A directive that tenders must be submitted without qualifications.
9 Advice and instructions with regard to alternative tenders (see page 45).
10 A statement regarding haul routes, that the contractor's vehicles should, or should not, use.
11 A reminder of the need to phase the work so as to avoid existing mains and services, unless permission is obtained from the appropriate authorities.
12 Notification that the amount of the successful tender and the tenderer's name will be made public, and directions as to the steps the tenderer should take if he objects to such procedure. (This will only apply in the case of public sector contracts.)
13 Directions as to how rates and extensions should be set out in the bill of quantities.
14 An indication of the period between the date of the award of the contract and its commencement.
15 A reminder that the employer is not bound to accept the lowest or any of the submitted tenders.

It should be noted that under Clause 11(2) of the *ICE Conditions of Contract* the contractor will be deemed to have inspected and examined the site, its surroundings and information available in connection therewith, and under Subclause (3) to have:

(a) based his tender on such available information, and on that inspection and examination, and
(b) satisfied himself *before submitting his tender* as to the correctness and sufficiency of his tender rates and prices.

The latter rates and prices constitute his 'offer' (see Chapter 1).

Enquiry
Invitations to tender are usually set out on a standard enquiry form to ensure that the returned tenders can be strictly compared. The time allowed for pricing varies depending on the complexity and size of the contract, the need for tenderers to obtain subcontractors' and suppliers' quotations, the employer, and the degree of urgency for an early start to the works. The usual minimum tendering period is four working weeks, with six weeks being about the average.

If a tenderer discovers variations between the various contract documents that render them ambiguous, or if the need for clarification of certain items arises, he should contact the engineer immediately so that such points can be examined and adjusted (if necessary) early in the tender period. If the documents require amendment, all tenderers must be informed and, if necessary, the tender period extended. Such notification should be in writing, but tenderers may be given advanced warning of the change by telephone if the circumstances so warrant.

Checking and appraisal

After tenderers have priced the bill of quantities and completed the form of tender, the documents are returned to the engineer by the given time and date. Late submissions are not normally accepted, and should be returned unopened to the tenderer. If a tenderer so wishes, he may, under English law, withdraw his offer at any time prior to the acceptance (see Chapter 1).

The returned tenders are compared and generally the three lowest bills of quantities then examined and checked in detail as follows:

1 Arithmetically, for mistakes in extensions and totals, etc.
2 Technically, for missing rates or incorrect interpretation of the bill layout.
3 Financially, for extraordinary rates and prices.

The checking procedure may reveal errors and discrepancies in the lowest tender which need to be resolved before the contract can be awarded. The method of dealing with errors and discrepancies will vary depending on the standard form under which the contract is prepared.

If the contract is of the 'lump sum' type where the offer is to execute the whole of the work for a contract sum or total, subject only to adjustment for ordered variations, price fluctuations and claims, the recognized procedure is to allow one of two alternative methods. The choice of method is made prior to tender and the tenderers notified accordingly. The alternatives are:

1 No corrections are permitted. The contractor must stand by the tender figures or withdraw his offer. If he withdraws, the second lowest tenderer is offered the job, subject to the above procedure.
2 Corrections are permitted. The contractor may confirm his offer, or he may amend genuine errors. If he chooses the latter he risks losing the job to the next lowest tenderer.

If the contract is carried out in accordance with the *ICE Conditions of Contract* (sixth edition) the total in the bill of quantities is the 'tender total', a figure used for tendering purposes only, and not used as the basis of the final account as is the case with the 'lump sum' type of contract. Under the sixth edition, the rates and prices predominate, and any arithmetical errors in extensions will automatically be corrected or eliminated when the works are remeasured and valued for the final account. Nevertheless, arithmetical errors need to be corrected to be able to compare the lowest tenders. The engineer should note any rates that are unusually high or low, and any items that have not been priced, and draw them to the tenderer's attention. Normally, contractors will be expected to stand by their low rates or withdraw their tenders and they will argue that high rates are 'their rates for the job' and will not expect to be asked to reduce them. However, if a contractor has clearly inserted a rate in error he may be allowed to change the adjustment item to accommodate the error, although such an action may increase his tender above the second lowest and so lose him the contract. Likewise, 'loaded' rates for items of a general or preliminary nature aimed at obtaining early money on the contract may be adjusted at the instigation of the engineer to spread the cost to the employer more evenly over the whole contract period.

Negotiations

After checking, and dealing with any errors and discrepancies, the lowest tender will normally be recommended by the engineer for acceptance.

If that tender exceeds the estimated total amount, or the employer's budget, it may be necessary for the employer and the engineer to negotiate with the tenderer to adapt the scheme to reduce the tender to match the estimate or budget. The next lowest tenderer should only be approached if the negotiations fail.

The final selection of a contractor may be held up until negotiations have produced a mutually satisfactory approach to the contract. If difficulties and problems between the prospective parties are not overcome, the lowest tender will be dropped and replaced by the second lowest, and negotiations will then start with that contractor.

If negotiations are protracted, they may delay the award of the contract beyond a reasonable time or beyond (if there is one) the stated period for acceptance. Where this occurs, the contractor may become entitled to revised rates and terms of contract.

All negotiations should be minuted and recorded and, where applicable, agreements reached appended to the contract documents.

Letter of intent

Sometimes, where it is necessary to get the works quickly under way or where there is likely to be a delay in finalizing the agreement, it may be necessary to notify the contractor that he has been successful so that he may recruit staff or make financial arrangements. This is done by means of a 'Letter of intent', which advises the contractor of the employer's intention to place the work with him, but the letter is not legally binding in the United Kingdom. The contractor should not commence work therefore, or incur costs, as he may find that they are irrecoverable. He should wait for the 'Letter of acceptance' before starting with the work.

Exceptionally to the foregoing, letters of intent sometimes give authorization to commence specific works such as site clearance and earthworks, and to incur expenditure up to a set limit. Where this is the case, legitimately incurred costs up to the set limit will be reimbursed.

Letter of acceptance

The 'Letter of acceptance' should be unconditional (see Chapter 1). It will usually be prepared by the engineer, but the acceptance will be made and signed by the employer (who is one of the parties to the contract). However, the engineer may 'accept' if he is given express authority to do so as the employer's agent.

The precise wording of the letter may vary but, for ICE sixth edition contracts, should include the words:

is accepted at the rates and prices contained in the bill of quantities totalling £. . . .

or

the revised tender total is made up as follows . . . .

References should be made in the letter to any negotiated agreements, commencement and completion dates, and all other relevant information. Should any new terms, conditions or items be added they will destroy the original 'offer' and change the 'acceptance' into a 'counter-offer', which will require an 'acceptance' by the contractor.

Once an offer has been accepted all the tenderers should be notified and provided with a list of tender figures.

Alternative tenders

Tenders should be free from qualifications. However, some contracts allow for alternative tenders to be submitted provided they comply with certain requirements. Others specifically provide a built-in alternative method of construction, so that the contract includes two types of design. For example, motorway and trunk road schemes falling under the auspices of the Department of Transport include both flexible and rigid forms of construction with each form being itemized separately in the bill of quantities. In such cases, the general summary in the bill of quantities has two totals, one for each type of construction, and the final decision as to which construction will be employed is deferred until after the contract is let.

Alternative tenders should be strictly in accordance with the original tender requirements. Where they involve modifications to design, the contractor should ascertain from the engineer what special design criteria and requirements apply. Generally, the following procedure should be adopted when contractors wish to submit alternative tenders.

1 The contractor should notify the engineer as early as possible during the tender period of any modified design proposals. Usually, there is a latest time limit, for example not later than two weeks before tenders are due to be returned. At this stage, the engineer may give a preliminary (but not binding) decision about the engineering acceptability of the proposals.
2 Alternative tenders must include supporting information, with drawings, calculations and a priced bill of quantities, to enable all aspects of the proposal to be evaluated.
3 Before final approval, the engineer may seek an independent check, the costs of which must be borne by the contractor.

4 If the alternative design and tender is accepted, the engineer will become responsible for the design.
5 Acceptance of the alternative will depend on the overall savings that the proposal will bring about. This includes project time and associated engineering and administrative costs.
6 In awarding the contract consideration will be given to the price tendered, to the engineering acceptability of the design, and to the various benefits, or otherwise, that may result from the alternative. In this last respect consideration will be given to:
   (a) Any financial saving on the lowest tender.
   (b) The effect on starting and completing the contract.
   (c) The cost of maintenance.
   (d) The costs and delays associated with the acquisition of additional land, etc. (if required).
   (e) The additional costs in assessing and implementing the alternative design.

Contractors' reasons for submitting alternative tenders vary, but clearly, are likely to result in some financial benefit for themselves. They may, for instance possess plant and equipment or have a particular expertise to suit their proposals. Additionally, they may have beneficial arrangements with certain specialist subcontractors. Their alternatives may be extensive, for example underpass walls comprising contiguous piles faced with gunite in lieu of the engineer's designed reinforced concrete, or minor amendments as shown in Figure 3.1.

In most cases, contractors must offer substantial cost savings to the employer for their alternative tenders to be accepted. Tenderers submitting alternative design proposals and tenders are usually required, also, to price the bill of quantities for the original scheme.

Two-stage tendering

Two-stage tendering is a competitive form of negotiated tendering used when the drawings have not been finalized and where there is only a simple basic or provisional specification. It takes several forms, the two most common of which are as follows:

First form

Stage 1
Selected contractors, following discussions with the engineer on how the proposed work should be carried out, are invited to tender by pricing a notional bill of quantities, a bill with approximate quantities or, maybe, a bill of quantities of a previous similar project. The returned tenders are required to include the contractors' proposals in sufficient detail to show how the work will be carried out.

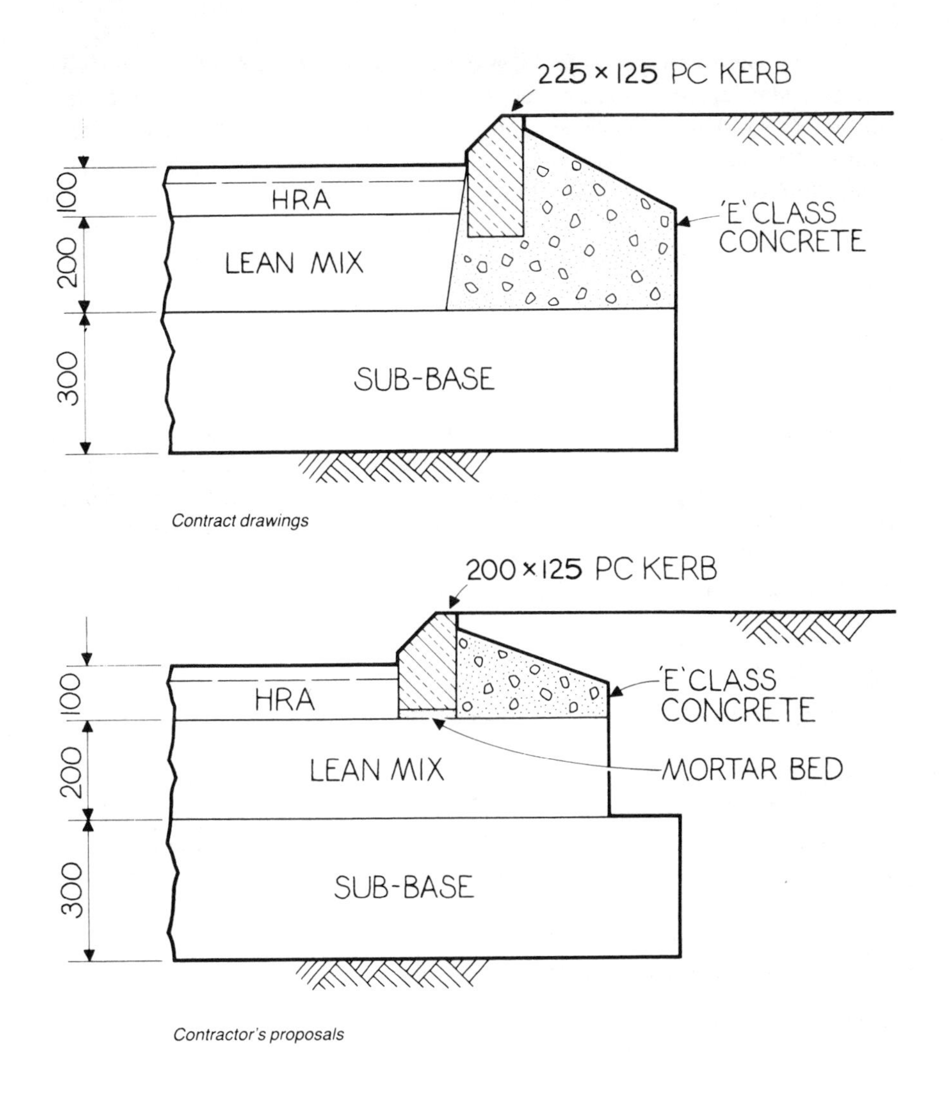

Figure 3.1 *Contractor's alternative proposals – amended road and kerb design*

Stage 2

After examination of the tenders and proposals, the most favoured contractor is selected. He is then required to work closely with the engineer to produce an economic design and a satisfactory programme. Subsequently, he submits a revised tender based on the final design, using the rates and prices in his original tender.

A variation of the second stage is that the most favoured contractor's proposal is selected following the return of tenders, and is then sent out to all the other contractors for them to price and submit new tenders. This second stage method is clearly less than satisfactory from the contractors' points of view since they are involved in a considerable amount of work in drawing up proposals and double pricing, all of which is likely to be abortive, and even if they are awarded the contract, the final proposals might not be theirs or meet with their approval.

Second form

Stage 1
The enquiry is sent to all the selected contractors who are given a limited time in which to examine the proposed scheme. They are not asked to tender for the work at this stage, but simply to attend a meeting after the examination to consider the proposed scheme, and to comment on, criticize and/or suggest alternative proposals. From the accepted suggestions and proposals a revised scheme is drawn up which may require further meetings with individual contractors to iron out details.

Stage 2
The revised scheme is sent to all the selected contractors for them to price.

Subcontractors

Civil engineering subcontractors fall into two categories: nominated and domestic.

Nominated subcontractors
A nominated subcontractor is defined in Clause 1(m) as:

> any merchant tradesman specialist or other person firm or company nominated in accordance with the contract to be employed by the contractor for the execution of work or supply of goods materials or services for which a Prime Cost has been inserted in the contract or ordered by the engineer to be employed by the contractor to execute work or supply goods materials or services under a Provisional Sum.

A nominated subcontractor, therefore, is usually a specialist subcontractor or supplier who is selected by the engineer. The engineer then instructs the main contractor to place an order with that preferred subcontractor. A 'prime cost item' means an item which contains (either wholly or in part) a sum which *will* be used for the execution of work, or for the supply of goods, materials or services (Clause 1(1)(k)). Subcontract work can also be carried out under a 'provisional sum', usually in cases where insufficient information was available to allow the work to be fully described prior to tender. Clause 1(1)(l) defines a provisional sum as a sum included in the contract for the execution of the work, or the supply of goods, materials or services, which sum *may* be used in whole or in part, or not at all, at the direction and discretion of the engineer.

Although the engineer has the choice, the final selection of the nominated subcontractor can be said to rest with the main contractor for Clause 59(1) states that the contractor shall not be under any obligation to enter into any subcontract with a nominated subcontractor against whom he has reasonable objection.

The procedure leading to selection of a nominated subcontractor is that prior to letting the main contract, or shortly after it has been let, the engineer will approach a number of subcontractors engaged in the type of specialist work required. They will be provided with the necessary documents to competitively tender for the work in a similar manner to that described under 'Single stage tendering'. The lowest tender will normally be preferred and the contractor who produced that tender will become the nominated subcontractor, unless the main contractor objects as stated above. Under those circumstances, the second lowest tenderer is likely to be selected.

Clause 59(2) sets out the procedure following the contractor's valid objection to a proposed nominated subcontractor or, if, during the course of the nominated subcontract, the contractor legitimately terminates the nominated subcontractor's employment owing to the latter not complying with the requirements of Subclause (1).

Under such circumstances the following options are available to the engineer:

1 Nominate an alternative subcontractor.
2 Vary the work, etc.
3 Omit the work, etc. so that it may be carried out by the employer himself, or his directly employed subcontractor.
4 Instruct the contractor to find his own domestic subcontractor to do the work.
5 Invite the contractor to do the work.

Subclause (3) makes the contractor responsible for the nominated subcontractor in all respects, except for the nominated subcontractor's design and specification etc.

Subclause (4) outlines the procedure to be followed should the nominated subcontractor be in default under the forfeiture clause. In this case, the contractor may terminate the nominated subcontract. However, the nominated subcontract cannot be terminated without the engineer's written consent. If the consent is withheld, the contractor will be entitled to receive engineer's instructions in accordance with Clause 13. If, on the other hand, the nominated subcontract is terminated, the engineer has the above five Subclause (2) options available.

The contractor must, having with the engineer's consent terminated the nominated subcontract, take all available steps to recover the additional expenses incurred under the subcontract (including the employer's expenses resulting from the termination). In the event of the contractor failing to recover all his reasonable expenses of completing the subcontract works as well as all his valid additional termination expenses, he will be reimbursed by the employer.

The valuation of nominated subcontractor's work falls under Clause 59(5), (6) and (7), and is dealt with in Chapter 7.

Domestic subcontractors

Domestic subcontractors are contractors who carry out work that the contract envisages being executed by the main contractor. However, for a variety of reasons, the main contractor may elect to sub-let portions of that work to other parties. If he does so, he must notify the engineer in writing in accordance with Clause 4 (see Chapter 4), unless the subcontractors are of the 'labour only' variety. The choice of domestic subcontractors and the method of selection rests with the main contractor. His method may be one of expediency, but is more likely to depend on the lowest price that is quoted by, or negotiated with, the various subcontractors. Although this sometimes involves unsavoury 'horse-trading', it can result in competitive prices and keen tendering for the main contract.

The valuation of domestic contractors' work is dealt with in Chapter 7.

4 Functions, responsibilities and obligations of the parties

Reference ICE Clauses: 1, 2, 3, 4 and 8, and the ICE Guidance Note 2A: Functions of the Engineer under the ICE Conditions of Contract

The rights, responsibilities and obligations of the employer, engineer and contractor fall into two categories. First, there are the specific duties and obligations required by the contract. These are set out in the contract documents. Then there are the general legal requirements, either of common or statute law. Some of both categories are better dealt with under specific headings, and as a consequence are discussed in following chapters. Thus, for example, the obligations of the employer relating to the possession of the site are to be found in Chapter 5 under that particular heading, whereas the more general obligations of the parties are examined in this chapter.

The employer

The employer is identified in the contract under Clause 1(1)(a) and includes the employer's personal representatives, successors and permitted assigns.

The employer is the client or promoter, and it is he who commissions the work and pays the design and construction costs. The employer is, therefore, the most important party to the contract, for without him there would be no contract and, therefore, no work for either the engineer or the contractor. Once the contract has been signed, however, the employer and the contractor become equal parties to the contract, each with duties and obligations one to the other, the most important of which are set out in the form of agreement (see Chapter 2). From the contractor's viewpoint, the principal obligation of the employer must be his undertaking to pay for the work done. Most of the employer's other obligations fall on his agent, the engineer.

It can be said, therefore, that the function of the employer is to provide the work and, subsequently, to finance that work.

Employers come from one of two sectors: the public sector or the private sector. The public sector includes government departments, local authorities, nationalized industries and public utilities, etc., and provides between 80 and 90 per cent of all civil engineering work in the United Kingdom. This is an excellent situation, as contractors are unlikely to find that, having done the work, they do not get paid because the employer has insufficient funds or has gone out of business. However, contractors relying solely on one type of work, and on one particular employer, are likely to get their fingers burned in times of Government restrictions and financial cutbacks. Public bodies deal with public money and must be seen to be spending that money both wisely and honestly. Consequently, the letting of their contracts normally follows some form of competitive tendering.

The private sector includes multinational organizations and corporations, and provides the civil engineering industry with only some 10 to 20 per cent of its work. Much of this work is of a highly specialized nature: the principal employers being petrochemical companies requiring refineries, gas and oil pipelines, docks and terminals, etc., or chemical works. Such bodies are entirely free to let construction work based on a variety of considerations and need not necessarily invite competitive tenders if the situation so warrants.

The engineer, the engineer's representative and the resident engineer

The engineer is identified in the contract under Clause 1(1)(c), but may be changed from time to time provided the employer notifies the contractor in writing. Clause 1(1)(d) simply defines the engineer's representative as a person notified as such from time to time by the engineer, and Clause 2(3)(a) requires that the contractor be notified in writing of the appointment.

The engineer is employed by the client to plan and design the project, to draw up the contract (generally including the preparation of the bill of quantities), to obtain tenders, to let and supervise the work, to authorize payments and to issue certificates. This is a similar role to that of the architect in building, although the latter's role is somewhat limited by comparison, thus reflecting the less risky nature of building work.

The architect has general, periodic supervision of the work, but the every day site supervision falls on the employer, and the contractor who is required to keep upon the works a 'competent person in charge'. The employer may appoint a clerk of works whose duty is solely to act as inspector on his behalf but under the direction of the architect.

Furthermore, in building the design and financial functions are shared between the architect and the quantity surveyor, whereas the civil engineer retains total responsibility for civil engineering contracts (see Figure 4.1).

Since the engineer acts on behalf of the employer, he is the agent of the employer (the principal), and as such his conduct is regulated by the laws of agency. The engineer must act honestly and obediently and exercise reasonable skill and care; and acts of the engineer, falling within the scope of his authority,

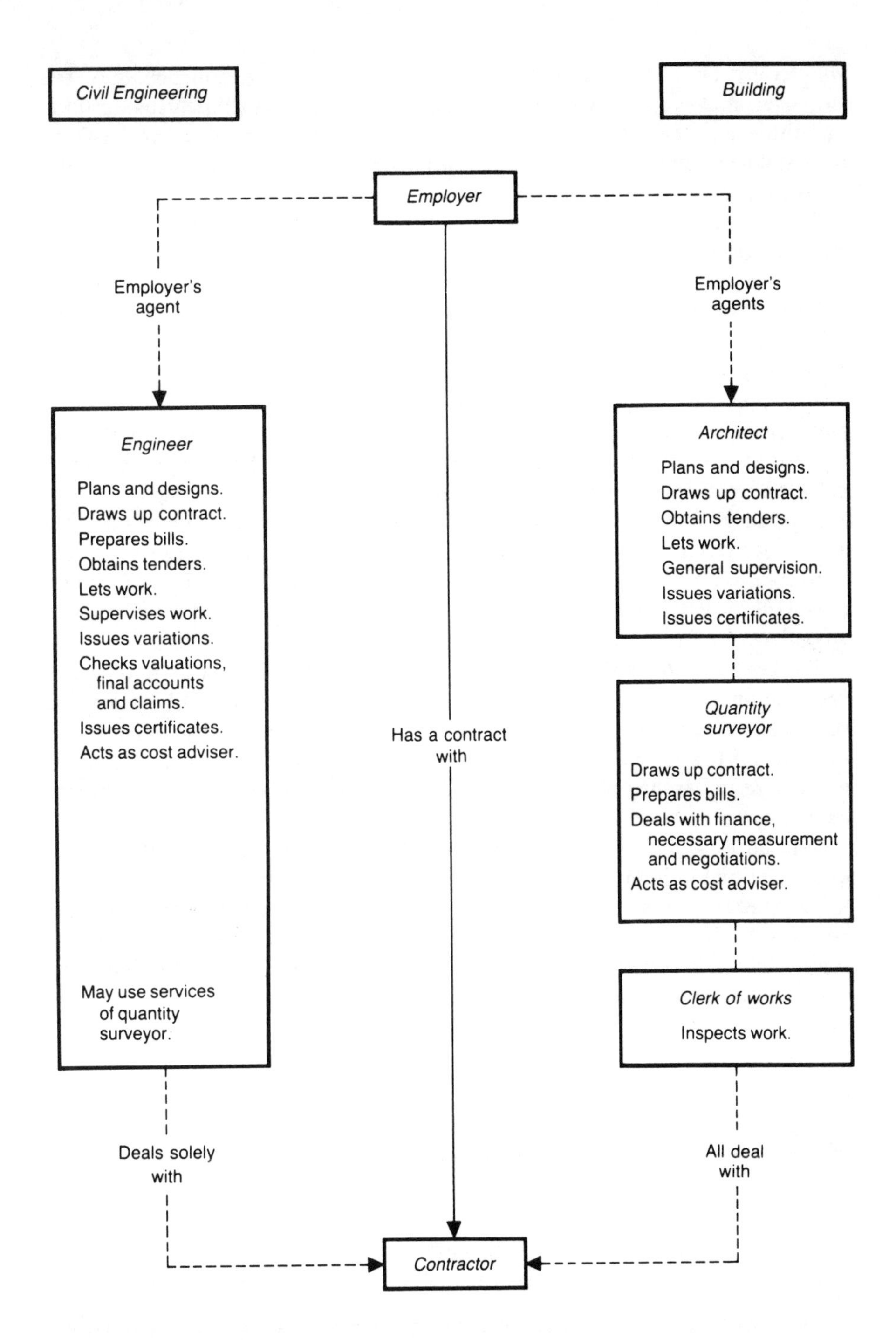

Figure 4.1 *Comparative roles of the civil engineer and the architect*

will legally bind the employer. Once the engineer, acting on behalf of the employer, makes a contract with a third party (the contractor) the usual result is that the engineer drops out, thus leaving the contract enforceable only between the employer and the contractor. The employer can then sue, or be sued, on the contract if the need arises.

The ICE requirements in relation to the engineer and his representative are set out in Clause 2.

Subclause (1) states that the engineer shall carry out the duties, and may exercise the authority, specified in or implied from the contract. Where the special approval of the employer is required before exercising such authority, he must comply with the requirements set out in the appendix.

The subclause concludes by stating that the engineer has no authority to amend the terms and conditions of the contract or to relieve the contractor of any of his contractual obligations, except as stated in the contract. (This latter exception is usually only in regard to ordered variations under Clause 51. See Chapter 7).

Subclause (2) requires the responsibilities of the engineer to be undertaken by a single, named chartered engineer, but not necessarily a chartered 'civil' engineer. If an engineer as above is not appointed as required under Clause 1, the engineer must nominate a chartered engineer and the contractor be notified accordingly in writing within 7 days of the contract being awarded to the contractor.

The responsibilities of the engineer's representative are primarily to the engineer and are set out in Subclause (3), which also states that he must watch and supervise the construction and completion of the works. The subclause goes on to say that he has no authority to relieve the contractor of any of his duties or obligations under the contract, nor to order any work involving delay or any extra payment, nor to make any variation of, or in, the works.

Subclause (4) makes provision for the engineer, as required, to give written authorization for the engineer's representative (or any other person responsible to the engineer) to act on behalf of the engineer. Notice of such authorization must be given to the contractor in writing, and will continue in force until the contractor is similarly notifed to the contrary. The engineer may not, however, delegate the powers given to him under the following clauses:

Clause 12(6) – delay and extra cost due to adverse physical conditions and artificial obstructions.
Clause 44 – extension of time for completion.
Clause 46(3) – provision for accelerated completion.
Clause 48 – certificate of substantial completion.
Clause 60(4) – certification of the final account.
Clause 61 – defects correction certificate.
Clause 63 – determination of the contractor's employment.
Clause 66 – settlement of disputes – arbitration.

Under Subclause (5) provision is made for the engineer, or his representative, to appoint any number of assistants to the engineer's representative. Such persons' names and their duties and scope of authority must be notified to the contractor. Those persons have no power beyond the powers necessary to discharge their prescribed duties, and those necessary to secure acceptance of materials and workmanship as being in accordance with the contract. Any instructions accordingly given are deemed to be instructions given by the engineer's representative.

The fifth subclause also covers situations where the contractor is dissatisfied with instructions of assistants of the engineer's representative. In such cases, the contractor is entitled to refer the matter to the engineer's representative for a decision. Similarly, under Subclause (7), if the contractor is dissatisfied with any act or instruction of the engineer's representative, he is entitled to refer the matter to the engineer for his decision.

Instructions given by the engineer or his representative are dealt with in Subclause (6), and must be in writing. However, if it is considered necessary, for instance to avoid delaying a critical operation, an instruction may be given orally. Such an instruction must be confirmed in writing as soon as possible, and the contractor must comply with both written and oral instructions.

This subclause covers situations involving CVIs ('confirmation of verbal instructions') whereby the contractor confirms in writing instructions given to him verbally. If the engineer does not contradict the contractor's confirmation in writing 'forthwith', that confirmation is deemed to be an engineer's instruction. The engineer (or his representative) must not, therefore, treat confirmations of verbal instructions lightly, and should reply to the contractor, recording any contradictions without delay.

Subclause (6) concludes by stating that the contractor has the right to be informed in writing as to which of the engineer's, or his representative's, duties or authorities cover their instructions given.

The final subclause requires the engineer to act impartially.

The engineer's representative on site is the resident engineer, who may or may not be aided by assistant engineers, depending on the size and complexity of the works. Additionally, the resident engineer will have the services of inspectors and, possibly, quantity surveyors. Some contracts, particularly motorway projects, often also have laboratory assistants on site to test concrete cubes and bituminous samples, etc.

The resident engineer is responsible for the day-to-day site supervision, ensuring that the work is carried out in accordance with the conditions, drawings and specification. His specific duties have to be authorized as required by Clause 2, but generally he will have authority to issue site instructions and variation orders, check interim valuations and recommend amounts for payment, provisionally agree new and additional rates, and advise the engineer on progress, delays and claims. He should also keep all records, registers, drawings, charts and correspondence that are necessary for the successful completion and historical record of the contract operations, bearing in mind that too much information is invariably better than too little. Additionally, the resident engineer is frequently required to deal with members of the public, the police and other persons in authority. In these situations he is often called upon to act as peacemaker or honest broker, and as such needs the forbearance and skills of a good public relations person. In his dealings with the contractor, the resident engineer needs to be positive and firm, but not inflexible.

In addition to the ICE requirements in relation to the engineer's representative, the Institution of Civil Engineers, following expressions of concern as to the role of the engineer and doubts as to the engineer's impartiality in matters relating to his own performance, drew up guide-lines for engineers. These were issued in 1977 under the title of *Guidance Note 2A: Functions of the Engineer under the ICE Conditions of Contract.*

The following are the principal points set out in the guidance note.

Functions of the engineer

The engineer is appointed by the employer to administer the contract, and the powers and duties vested in him are restricted to him.

The engineer's duty as agent of the employer is to supervise the construction work to ensure that the contractor complies with the contract, but of at least equal importance are his functions of decision-making relating to the allocation of responsibility for risks, changes in construction and time.

The engineer is expected to act 'impartially, honestly and with professional integrity, towards both parties to the contract'.

The engineer in relation to the rights and obligations undertaken by the employer and the contractor

It is acknowledged that it is impossible for either of the parties to a contract to foresee every eventuality likely to affect the construction, and points out that the contractor is responsible for risks that could reasonably be foreseen by an experienced contractor. Where unforeseeable risks occur they must be considered separately, which may lead to extra costs and extensions of time.

The above allocation of risks is said to be in the interests of both parties, for if the contractor were obliged to accept all the risks he would have to estimate for too many unknowns.

Provision for dealing with the risks relating to unforeseeable circumstances is made in the contract thus giving the engineer power and duty to exercise his professional judgement to make impartial decisions in such matters as:

1 The contractor's reasonable risks (Clause 12)
2 Extension of time (Clause 44)
3 The valuation of changes to the works (Clause 52)
4 Differences between the employer and the contractor (Clause 66)

In carrying out his functions the engineer must act impartially and ought to inform the employer of:

1 Progress
2 Performance
3 Variations
4 All other matters in which the employer has a legitimate interest

The engineer must consider any representations of either the employer or the contractor, be free to consult and seek advice on any matters before reaching decisions, and he ought to record the principles forming the basis for his decisions for future reference.

The impartial engineer
The engineer's unique position enables him to deal with circumstances which may arise during the construction. Such circumstances should be resolved expeditiously, and in such a way as to maintain satisfactory progress of the construction work. The trust of both parties in the impartiality of the engineer will have a beneficial effect on site relationships and contribute to the successful completion of the contract.

If the engineer fails to act impartially, it will enable the contractor to challenge all decisions made by the engineer, and may be grounds for him suing the employer for breach of contract.

The contractor and the contractor's agent

The contractor is identified in the contract under Clause 1(1)(b) as the person or persons, firm or company to whom the contract has been awarded by the employer, and includes the contractor's personal representatives, successors and permitted assigns.

Under the contract, the contractor undertakes to carry out the work in accordance with the contract, specifically to construct and complete the works. This undertaking is set out in the form of agreement, and repeated in Clause 8. Additionally, Clause 8 says that the contractor must take full responsibility for the adequacy, stability and safety of all site operations and methods of construction, and provide all labour, materials, equipment, temporary works and transport; indeed everything necessary to carry out and complete the works. Furthermore, as has been pointed out earlier in this chapter, the contractor is responsible for the contract risks that could reasonably be foreseen by an experienced contractor.

The contractor is not responsible for the design or specification of the permanent works, nor for any temporary works designed by the engineer, but is responsible for, and must exercise all reasonable skill, care and diligence in designing, work where the contract expressly provides for permanent works to be designed by him.

Under Clause 3, neither the employer nor the contractor is permitted to assign (transfer to another party) either the whole, or any part of the contract, or any of their benefits or interests under the contract without the written consent of the other party. Such consent should not unreasonably be withheld.

The contractor is not allowed to sub-let the 'whole' of the works without the employer's written prior consent. The contractor may sub-let, however, parts of the works or design provided he has the consent of the engineer, but consent requires written notification of the extent of the work or design and the name and address of the subcontractor prior to entry on to the site, or of the designer prior to his appointment. Notification is not required, however, for the employment of labour-only subcontractors.

Under Subclause (5), the engineer may order, following a written warning, that the subcontractor be removed from the site if the subcontractor:

mis-conducts himself,
is incompetent, or
is negligent

in performing his duties, or fails to conform with any particular safety provisions set out in the contract, or persists in conduct prejudicial to safety or health. Such subcontractor may not be employed again upon the works without the engineer's written permission.

Contractors fall into various general categories. First, there are those who operate on a national or international level. By definition this type of company tends to be large, although it need not necessarily be so, particularly if it is of a specialist nature, for example temporary steel bridge, specialist equipment or services contractor. Companies in this category may also be subsidiary businesses of large conglomerates or, particularly in the case of companies operating overseas, offshoots of well-known parent companies.

Those operating abroad often combine with other European companies that may be specialists in particular types of work, to form consortiums. Besides the advantages of additional resources and expertise, such an arrangement often has the benefit of surmounting language difficulties. For instance a British company combining with a French company for work in French-speaking North Africa. They may also team-up with local companies (a necessary requirement in some Arab countries) which has distinct advantages in that it eases problems related to language and local customs, simplifies the acquisition of local resources and payment of labour, and is a means of overcoming obstacles presented by local officials.

Generally, companies in this category are well organized, highly competitive, extremely cost conscious, have efficiently run sites with established procedures supported by good head office back-up.

Second, there are contractors who operate on a regional or local level, and only undertake projects of a particular type or size. They tend to operate within a limited radius of their base and normally take on work that does not exceed a set financial limit. They are frequently both building and civil engineering contractors combined, unlike the previous type which have a tendency to specialize solely in building or civil engineering work. Some firms, on the other hand, are of a specialist nature and they may operate as general contractors on jobs of their special type, for example large earthworks, asphalt road surfacing, thrust boring, etc. On particularly large contracts these specialist contractors may even sub-let general work ('builders work') to contractors who, in other circumstances, would be main contractors.

Third, there are contractors who design and construct the works. The manner in which this type of contractor operates is described in Chapter 2.

The contractor's equivalent or counterpart to the resident engineer is the agent, who may also be known as the project manager, and who for the purposes of the contract is deemed to be the contractor. The agent is responsible for the efficient execution of the site operations, normally being directly responsible to the contracts manager. He controls the contractor's site operations; being especially concerned with planning and progress; the provision and deployment of labour, materials and equipment, and the subsequent transfer or disposal of those surplus to requirements or no longer required; site organization; labour relations; and dealing with the engineer's representative, the public and any local or other officials.

Normally, the agent will have the services of a quantity surveyor to deal with measurement and related site records, the preparation of interim valuations, claims and final accounts, and the assessment of payment to subcontractors.

It should be stressed that the principal object of all construction contracts is the successful completion of the work, resulting in the employer obtaining that which he is paying for. Ideally he will get this at the earliest possible time, with the least possible trouble and to the highest possible standard within the provisions of the contract.

This can only come about if all the parties involved aim to work together – not against each other as is the case on occasions. Co-operation, despite differences of opinion and personality, is essential.

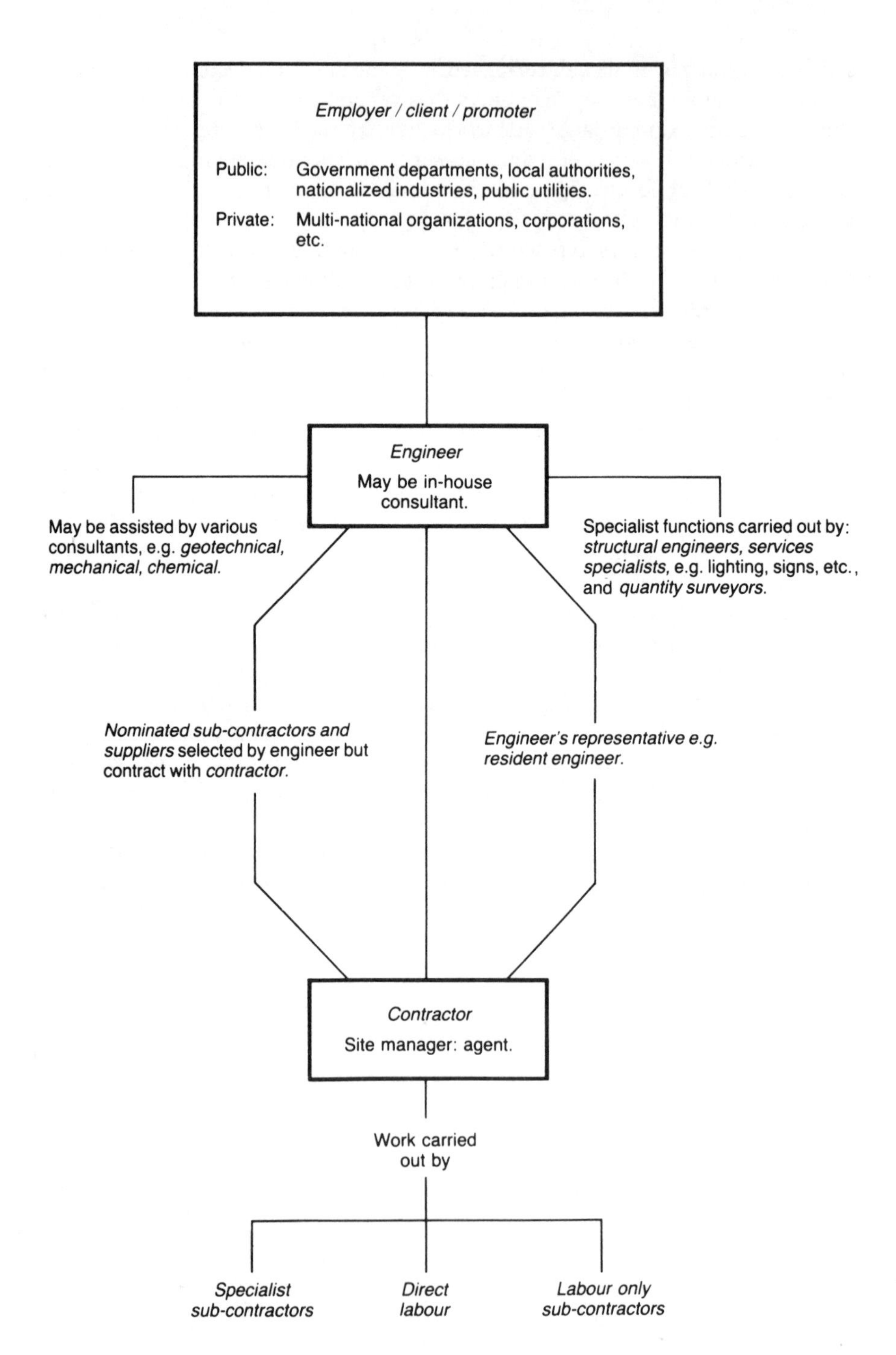

Figure 4.2 *Civil engineering relationships*

5 Possession of the site, completion and extensions of time

Reference ICE Clauses: 41–46 inclusive, 48 and 49

In addition to the express terms relating to the possession of the site that are set out in the contract, there are also general requirements of law. These are concerned with occupiers, who may be employers, contractors, or even subcontractors, and with employees and visitors to, or persons affected by, the premises. The general requirements fall under both common and statute law and are dealt with first, under general legal requirements.

General legal requirements

Statutory requirements associated with the possession of the site are generally concerned with negligence, and make frequent reference to the 'occupier'. Occupier in this context has a rather different meaning from that generally used. The occupier is a person having some degree of control over the premises. He may be the owner or the tenant; or on construction sites, the contractor, or even the employer and contractor together. Likewise, a subcontractor may also be the occupier of the site, or part of it.

Under common law, the occupier of a premises has a duty to protect certain categories of visitors from the dangers of the premises. That duty was widened in scope by the Occupiers' Liability Act 1957, so that the occupier of a premises owes a duty of care to all lawful visitors on that premises, unless the duty is modified or excluded by agreement, subject to the restrictions on so doing under the Unfair Contract Terms Act 1977. The occupier must exercise reasonable care for the safety of all persons using the premises for the purposes for which they are invited or permitted to be there. Premises include land, buildings and construction sites. They also include movable structures, such as vessels and vehicles, lifts and hoists, and fixed structures such as scaffolding.

In general, the duty of care does not apply to unlawful visitors, that is, trespassers. They enter premises at their own risk, although they may now be owed a duty of care under the Occupiers' Liability Act 1984. However, the occupier must not intentionally harm trespassers, for example by setting traps for them. Furthermore, the occupier has a duty to take such steps as common sense and humanity require to exclude trespassers from dangerous premises, to give them adequate warning of existing dangers, and to reduce the dangers. Failure to do so may result in the occupier being held liable for injury suffered by the trespasser.

As a general point of law, it should be noted that if an occupier of land takes on to and keeps upon that land things likely to cause damage if they escape, he must be particularly careful, for in the event of escape he is likely to be held liable for the consequences, even though he may not have acted negligently.

Once the contractor has taken possession of the site he has certain obligations to his employees. Prior to 1974, if the Factories Act 1961 or the Offices, Shops and Railway Premises Act 1966 had not been breached, an employee injured at work could sue his employer for damages for negligence under common law. Normally, this would have been a civil action and only very rarely a criminal prosecution. Further protection for workpeople was considered necessary to safeguard their health and to prevent accidents. This resulted in the Health and Safety at Work etc. Act 1974, which introduced regulations and codes of practice aimed at maintaining and improving existing standards. Breach of the Act or the regulations made under it is a criminal offence, although an employee who requires compensation for injuries suffered at work will still have to bring a civil action either for negligence or for breach of a statutory duty.

A further aim of the Act is to encourage safety consciousness not only in employers, but also in all those who are concerned with places of work, imposing general duties on employers and employees, the self-employed and persons controlling workplaces to ensure that working situations are safe for people working in them, visiting them, or otherwise affected by them.

Normally, for an action in negligence to be upheld, actual damage has to be suffered. Under the Health and Safety at Work etc. Act no injury need occur before the employer is prosecuted, and failure to provide adequate welfare facilities is a criminal offence which could result in prosecution.

Employers must take proper care, as far as is 'reasonably practicable', for the safety of all persons, including total strangers, who are affected by work performed on premises under their control. This includes premises where there are no employees, for example establishments using automatic machines such as 'fruit machines', washing machines and dry-cleaners, and the like.

Employers include managers and persons in control at any level. All levels of management should, therefore, have their health, safety and welfare duties clearly defined so that they are fully aware of the limits of their responsibilities. For their part, employees must take 'reasonable care' for their own health and safety, and also for that of others.

Employers are required to formally set out their safety rules and arrangements, and employees must be able to examine them. Employers should also as far as is reasonably practicable:

1 Provide and maintain safe plant and equipment, and safe systems of work.
2 Provide safe arrangements for the use, handling, storage and transportation of articles and substances.
3 Provide information, instruction, training and supervision.
4 Provide and maintain a safe working environment, free from health risk, with adequate facilities and arrangements for employees' welfare.

The employer must use 'all reasonable persuasion and propaganda' to persuade employees to comply with the safety requirements. Failure to do so could result in the employer being held responsible on the grounds of 'consent or connivance' in the offence.

The Act requires all employers to look after their own employees regardless of where they work, or the capacity in which they act. Responsibility for subcontractor's employees, therefore, falls on the subcontractor. However, the main contractor, as controller of the site, is responsible for their safety, and he must ensure that the site, materials, plant and equipment are all safe.

Possession of the site

Possession of the site is clearly linked with the commencement of the works Clause 41, which states that the date for commencement is:

1 the date specified in the appendix, or if no date is specified
2 a date within twenty-eight days of the engineer's written notification of the award of the contract, or
3 such other agreed date.

The contractor must commence the works on, or as soon as is reasonably practicable after the works commencement date. It is debatable as to how long is reasonable. The clause concludes by stating that the contractor shall proceed with the works with due expedition and without delay.

A point worth noting is that the date for commencement is the start of the period during which the works must be completed (see Clause 43). A late start by the contractor, therefore, or a delayed acceptance may affect the profitability of the contract. For the same reason if a long delay between the submission of tenders and acceptance is anticipated by the engineer and the employer, the tenderers should be informed at the tender stage.

Clause 42 deals with possession of the site, and access and states that the contract may prescribe:

1 the portions of the site to which the contractor is to be given possession, and their extent.
2 the order of availability of such portions.
3 the availability and nature of access provided by the employer.
4 the order in which the works are to be constructed.

Under Clause 42, the contractor must be given, on the works commencement date, possession of as much of the site and access thereto as may be required to enable the contractor to commence and proceed with the construction. From then on, the contractor is entitled to possession of, and access to, further portions of the site as may be required to enable him to proceed with the works in accordance with the programme accepted under Clause 14. If the contractor is delayed and/or incurs additional costs as a result of not being given possession of the site in accordance with this clause, the contractor is entitled to an extension of time, and to payment of such costs and profit as determined by the engineer. The engineer must notify both the contractor and the employer of the time and amount so determined.

Finally, Clause 42 requires the contractor to bear all costs and charges for any access that he requires additional to that provided by the employer. If the contractor requires facilities outside the site, he must meet the cost.

Completion of the works

In accordance with Clause 43, the whole works and any section so required must be substantially completed within the time for completion stated in the appendix to the form of tender, calculated from the works commencement date or any extended time that may have been allowed under Clause 44 (see following section).

Clauses 45 and 46 are relevant to the completion of the works. Clause 45 does not permit the contractor to work at night or on Sundays, that is, outside normal working hours, without the engineer's written consent unless it is:

1 Unavoidable,
2 Absolutely necessary for the saving of life or property, or
3 For the safety of the works

In the above circumstances the contractor must immediately advise the engineer.

Work that it is customary to carry out outside normal working hours, or by rotary or double shifts, is excluded from this clause. Should such work not be permitted, it must be clearly indicated in the contract documents so that tenderers can make due provision in their rates and prices.

Under Clause 46, the planned rate of progress is expected to be maintained. If for any reason (other than one entitling the contractor to an extension of time) the contractor, in the opinion of the engineer, falls far enough behind the programme to jeopardize substantial completion by the scheduled completion, or extended time for completion, the engineer must inform the contractor in writing. The contractor is then obliged, at his own cost, to take the necessary steps, subject to the engineer's consent, to redeem the situation. If this requires night or Sunday working, the engineer's permission must be sought and such permission should not be 'unreasonably refused'.

Subclause (3) of Clause 46 entitles either the employer or the engineer to request that the contractor completes the works, or any section of the works, within a shorter period than the agreed or extended time. If the contractor agrees to accelerate his programme, the terms of payment and conditions for so doing must be agreed beforehand.

When the contractor considers that the whole of the works, or any specified separate section, have been substantially completed and satisfactorily tested, he may notify the engineer in accordance with Clause 48. At the same time, the contractor must give a written undertaking to finish any outstanding work within the specified time(s) agreed, or if no such time(s) is agreed, as soon as practicable during the defects correction period (Clause 49(1)).

Clause 48 requires the engineer, within twenty-one days of the date of delivery of the notice of substantial completion, to:

1 Issue the certificate (with a copy to the employer), stating the date of substantial completion, or
2 Inform the contractor in writing of all the work required to be carried out before he will issue a certificate. When such work has been completed to the engineer's satisfaction, he should issue a certificate of substantial completion within twenty-one days.

Clause 48 also provides for the issue of certificates of substantial completion for:

1 Any substantial part of the works that has been occupied or used by the employer (other than provided for in the contract), following a written request from the contractor.
   The certificate becomes effective from the date of delivery of the request, and the contractor is deemed to have undertaken to complete any outstanding work during the defects correction period.
2 Any part of the works that has been substantially completed and satisfactorily tested before completion of the whole of the works.
   The contractor is obligated to complete any outstanding work as in 1 above.

Finally, the issue of a certificate of substantial completion for any section or part of the works before completion of the whole is not deemed to certify completion of any ground or surfaces requiring reinstatement, unless expressly stated.

Extension of time for completion

An extension of time for completion can be granted under Clause 44 in respect of the following:

1 Ordered variations
2 Increased quantities
3 Any other cause of delay referred to in the conditions
4 Exceptional adverse weather conditions
5 Other special circumstances of any kind whatsoever

If the contractor seeks an extension of time for substantial completion he must give the engineer full and detailed particulars within twenty-eight days after the cause of the delay having arisen, or as soon thereafter as is 'reasonable'.

Upon receipt, the engineer should make an assessment of the delay based on the particulars submitted and the circumstances known to him. In the absence of a claim for an extension of time from the contractor, the engineer may grant one, if he thinks fit. In both cases the contractor should be informed of the assessment in writing.

If the engineer considers that the delay 'fairly' entitles the contractor to an extension of time for substantial completion, an interim award should be granted immediately and, as with the assessment, the contractor should be notified as to whether or not time has been awarded and, if so, its extent.

Not later than fourteen days after the original or revised completion date, the engineer should review all the particulars and circumstances and make a final assessment. Both the employer and the contractor should be notified if the assessment does not entitle the contractor to an extension of time.

Within fourteen days of the issue of the certificate of substantial completion, the engineer should review all the circumstances and finally determine and certify to the contractor with a copy to the employer the overall extension of time (if any) to which he considers the contractor is entitled. The final review cannot decrease any extension of time already given.

It should be noted that Clause 44 is purely a time clause, and does not entitle the contractor to additional payment. The extension of time granted under this clause only relieves the contractor of liability for liquidated damages for the period of the extension. However, an extension of time may lead to successful claims for additional payment under other clauses.

6 Quality of workmanship and materials

Reference ICE Clauses: 4, 26, 36–40 inclusive, and 59

The contract requirements in relation to workmanship and materials are set out on the drawings and in the specification (see Chapter 2), and in the conditions of contract. The requirements of the drawings and specification are likely to include references to standards of quality and performance laid down by various British Standards, and Codes of Practice. In addition to the specific contractual demands, there are certain legal requirements regarding materials and workmanship that have to be met.

Statutory requirements

In civil engineering contracts, the contractor is usually required to give an express undertaking to comply with all statutory provisions, including by-laws. This, effectively, includes the Unfair Contract Terms Act 1977, the Sale of Goods Act 1979, and the Supply of Goods and Services Act 1982.

Clause 26(3) of the *ICE Conditions of Contract* states that the contractor must ascertain and conform in all respects with the provisions of any general or local Act of Parliament, and the regulations and by-laws of any local or other statutory authority, and with the rules and regulations of public bodies and companies whose property or rights are affected by the works, and must indemnify the employer against all penalties and liability of every kind for breach of such Act, regulation or by-law. There are three provisos to the above:

1 The contractor is not required to indemnify the employer against the consequences of any breach which is the unavoidable result of complying with the contract or engineer's instructions.
2 If the contract or engineer's instructions do not conform with any of the statutory requirements, the engineer must issue instructions and order a variation to ensure conformity with such Act, regulation or by-law.
3 The contractor is not responsible for obtaining planning permission for works designed by the engineer, and the employer warrants that all permissions have been, or will be, obtained in due time.

The sale or supply of goods and services is governed by numerous Acts of Parliament. Those most applicable to construction contracts include the Sale of Goods Act 1979, and the Supply of Goods and Services Act 1982.

Specific points of interest are that goods must be of merchantable quality if sold in the course of the seller's business. However, if the buyer was told of or, alternatively, should have seen any defect when he examined the goods, the vendor is protected. The buyer is not obliged to examine the goods prior to purchase. If he does so, however, the vendor is not liable for defects which should have been revealed by the examination. Neither is he liable for defects pointed out to the buyer prior to the sale.

Goods sold in the course of a business must be reasonably fit for the purpose for which they are required by the buyer. That fitness may apply to the container in addition to the goods.

Goods sold by description must correspond with the description. If the sale by description includes a sample, the goods must correspond with both the description and the sample.

Goods sold by sample must correspond in quality with the sample and the buyer must be given a reasonable opportunity to compare the goods with the sample.

Goods must be free from hidden defects that are likely to make them become unmerchantable later.

The above conditions may be varied or excluded from contracts of sale by an appropriately worded clause but, following the Unfair Contract Terms Act 1977, clauses excluding or restricting the implied rights of buyers are not permitted in the case of consumer sales. They are permitted in other sales, if considered 'reasonable' at the time the contract was made.

A consumer sale is defined under the 1977 Act as a sale (other than by auction or competitive tender) by the seller in the ordinary course of business where the goods are of a type usually bought:

1 For private use or consumption
2 By a person who is not himself a dealer buying in the course of business, that is, a person buying for private use.

The courts imply terms into other contracts if there is no agreement to the contrary. Thus, in contracts for building or civil engineering work there will be implied terms that work and materials are:

1 Of good quality, and
2 Reasonably fit for the purpose for which they are required, in so far as the employer relied on the skill and judgement of the contractor.

The 1982 Supply of Goods and Services Act (which does not apply to Scotland) also provides for the above terms to be implied in certain contracts for:

1 The transfer of ownership of goods, other than by direct sale

2 The hire of goods
3 The supply of a service

Clearly, the latter two are particularly applicable to the construction industry.

The Unfair Contract Terms Act 1977 applies to these contracts in the same way as it does to contracts for the sale of goods (with appropriate modifications) so as to prevent unreasonable use of exemption and limitation clauses. Additionally, under the 1977 Act, liability for the tort of negligence (e.g. under the Occupiers Liability Act – see Chapter 5) cannot be:

1 Excluded or restricted where negligence results in death or personal injury.
2 Excluded or reduced in the manufacture or distribution of consumer goods.

It should be noted that as far as construction contracts are concerned, private persons employing contractors are classified as consumers.

Under the Consumer Protection Act 1987, manufacturers of goods are strictly liable if their goods are defective and cause injury or damage.

ICE requirements: main contractors

The quality of materials and workmanship, together with tests, their cost, and samples are covered by Clause 36.

All materials and workmanship must be as described in the contract and be in accordance with the engineer's instructions. They must also be subjected to whatever tests the engineer directs. Regardless of the foregoing, however, the contractor is obliged to produce materials and workmanship in line with good building practice. The contractor must provide all assistance and equipment normally required for examining, measuring and testing work, together with all information about materials as required by the engineer, before the materials are incorporated in the works.

All samples required under the contract must be supplied by the contractor at his own cost. The cost of any not provided for in the contract must be met by the employer.

The cost of making tests must be borne by the contractor if such tests are:

1 Clearly intended or provided for in the contract and described in the specification or bill of quantities in sufficient detail to enable the contractor to have priced or allowed for them in his tender.
2 Other than as in (1) above, but produce results not in accordance with the requirements of the contract.

Under Clause 37 the engineer and persons authorized by him have right of access at all times to the works, site and workshops, etc., and to places whence materials, manufactured articles and machinery are being obtained.

Clause 38(1) deals with the examination of work before it is covered up. It states that no work is to be covered up without the consent of the engineer, and that the engineer is to be given full opportunity to examine and measure any work about to be covered up. The contractor must inform the engineer when such work is ready for examination, and the engineer must examine and measure the work without unreasonable delay, unless he considers it unnecessary and advises the contractor accordingly.

Clause 38(2) states that the contractor must uncover work previously covered up, if directed to do so by the engineer. The contractor must subsequently reinstate and make good the work to the satisfaction of the engineer. If the work is found to have been executed in accordance with the contract, and the contractor has complied with Clause 38(1), the cost of uncovering, reinstating and making good, etc. must be borne by the employer. In any other case, all the costs must be borne by the contractor.

Under Clause 39, the engineer has the power to instruct in writing:

1 The removal from site of any materials which, in the opinion of the engineer, are not in accordance with the contract.
2 The substitution with materials in accordance with the contract.
3 The removal and proper re-execution of any work in which the material, workmanship or contractor's design, in the opinion of the engineer, is not in accordance with the contract.

It should be noted that since there is no mention of any payment to the contractor, the above must be at his expense.

If the contractor defaults in carrying out such instruction the employer is entitled to employ and pay others to carry out the same, and all costs incurred must be borne by the contractor, either by payment to the employer, or by deduction from payments due to the contractor. Both parties should be notified of the amount that the engineer has determined as recoverable costs.

Should the engineer fail to disapprove any work or materials at any particular time he still has the power to subsequently disapprove such work or materials.

Clause 40 deals with suspension of work. It says that the contractor must, on the written orders of the engineer, suspend the whole or part of the works for such time and in such manner as the engineer considers necessary, and that the contractor must protect and secure the work. The contractor will be paid the extra cost incurred in so doing, unless the suspension is:

1 Provided for in the contract.
2 Due to weather conditions, or default on the part of the contractor.
3 Necessary for the proper execution of the work or for the safety of the works in as much as such necessity does not arise from some act of the engineer or employer, or from any of the 'excepted risks' (see Clause 20).

The engineer must take any delay caused by the suspension into account when determining an extension of time, unless the suspension is provided for in the contract, or results from some default on the part of the contractor.

If the suspension lasts for more than three months, the contractor may serve a written notice on the engineer requiring permission to recommence the work within twenty-eight days from receipt of the notice. If such permission is not given within the twenty-eight day period, the contractor may serve a further written notice on the engineer electing to:

1 Treat the suspended part of the works as an omission under Clause 51 (ordered variations) or,
2 Treat the suspended works (where it relates to the whole of the works) as an abandonment of the contract.

ICE requirements: subcontractors and suppliers

The contractor is normally responsible for work carried out by subcontractors.

Clause 4 allows the contractor to sub-let the whole of the works with the prior consent of the engineer, and to sub-let parts of the works or their design without his consent, although prior written notification and details of such are required. Notification of labour-only subcontractors, however, is not required. Clause 4 then goes on to state that the contractor shall be liable for all work subcontracted and for acts, defaults or neglects of any subcontractor, his agents, servants or workpeople. The clause also gives the engineer power to remove from the works any subcontractor who performs his duties incompetently or negligently. (See also Chapter 3.)

Responsibility for subcontractors also includes nominated subcontractors. Clause 59(3) of the conditions states that the contractor shall be responsible for the work executed, or goods, materials or services supplied by a nominated subcontractor 'as if he had himself executed such work or supplied such goods, materials or services'.

British Standards

British Standards are prepared and issued by the British Standards Institution. The same body also prepares and publishes Codes of Practice, and the *British Standards Yearbook* which lists and briefly summarizes all British Standards and Codes of Practice. The individual standards are prepared by committees of experts generally familiar with the subject matter of a particular standard. Thus, the most thorough investigations in line with current knowledge and technical development are undertaken prior to each standard being produced.

In all, there are some 11,000 operative British Standards covering a wide range of industries and subject matter. Their content may define or determine:

1 Terms, definitions and symbols.
2 Standards of quality and performance and dimensions.
3 Tests.

Their principal objective is to provide recognized national standards of quality of materials and components, and to define the tests with which they must conform. The value of British Standards is so well recognized that almost all specifications contain references to them. This is particularly so in civil engineering contracts where they are widely used. For instance, Part 7 of the Ministry of Transport's *Specification for Highway Works* (1986) used for motorway and highway schemes, published by HMSO, lists a total of 236 British Standards to which reference is specifically made.

By using standards in this way, the length of specification clauses is reduced and yet a good standard of product is ensured. Furthermore, it helps the contractor by standardizing specification clauses, thus reducing the contractor's need for examination of numerous descriptive clauses, often of considerable length and with widely varying requirements. Contractors, as a consequence, are likely to have a fair knowledge of the contents of clauses containing the more common standards and to know precisely what is required, can more speedily price the work and ultimately arrive at a more genuine comparative tender. Clearly, a sound understanding of the scope and contents of any particular standard is required before it should be included in a specification.

Codes of Practice

Codes of Practice, as mentioned previously, are prepared and issued by the British Standards Institution. They set out recognized good standards of practice. In all they total some 350 and many cover the design, construction and maintenance aspects of civil engineering work, tending to be extremely comprehensive and well illustrated.

As with British Standards, they often form an integral part of contracts, particularly specifications, and have resulted in a general all round improvement in standards of work. Their popularity in civil engineering, however, is somewhat less than in building, and an indication of the extent to which they are used is probably reflected by Part 7 of the *Specification for Highway Works* (1986) which makes reference to only five codes, although others are published under various British Standards included in the document.

7 Measurement and valuation of the works

Reference ICE Clauses: 2, 12, 51, 52, 55, 56, 57, 59 and 60, and the special conditions clause on contract price fluctuations

The measurement and valuation of the works is, clearly, a prerequisite to the certification and payment, which is described in Chapter 8.

Most contracts are of the bill of quantities or measurement type and their measurement and valuation is governed by Clauses 55–57 inclusive of the *ICE Conditions of Contract* which are described below. The means of assessing amounts due for payment to contractors in the case of other types of contract is dealt with in Chapter 2.

ICE general requirements

Clause 55 sets the scene for the measurement of contracts by stating that the quantities in the bill of quantities are not to be taken as the actual and correct quantities of the works, and that any error in description, or omission, shall not vitiate the contract or release the contractor from the execution of the works, or from any of his obligations or liabilities. Any error or omission must be corrected by the engineer, and the value ascertained in accordance with Clause 52 (see 'Variations to the contract'). Correction of errors and omissions does not include wrong estimates in descriptions, or wrong rates and prices inserted in the bill of quantities by the contractor.

Under Clause 56 responsibility for admeasurement and valuation of the work done in accordance with the contract rests with the engineer. Although the clause does not specifically say so, it must be assumed that responsibility is to the employer, since under Clause 60 the contractor has to produce interim and final accounts (which include his measurements) for the engineer to check.

Clause 56(2) is unique in that it allows the engineer to increase or decrease rates and prices in the bill of quantities. This situation arises if the actual quantities executed in respect of any item are greater or less than those stated in the bill of quantities to such an extent that, in the opinion of the engineer, the changed quantities render the original rates or prices 'unreasonable or inapplicable'. In these circumstances, the engineer must consult with the contractor before fixing an appropriate new rate, and having decided on a rate, notify the contractor accordingly. It is clearly debatable as to the point at which a rate becomes 'unreasonable or inapplicable', and depends on the particular item, the size of the original quantity, and the relevant circumstances, etc. However, as a general rule, changes would not be considered unless the change in quantity is in excess of 10 per cent of the bill of quantities figure.

Clause 56(3) instructs the contractor to assist the engineer in taking measurements whenever the engineer requires parts of the works to be measured and to furnish all particulars required for that purpose, providing the engineer gives 'reasonable notice' to the contractor of his requirements. If the contractor fails to attend to assist, the measurement made or approved by the engineer will be deemed to be the correct measurement of the work. In practice, engineers and contractors often have a sensible arrangement whereby they call on each other, sometimes at very short notice, to agree records or measurement of special items of work, particularly where breaking out and removal is involved, or work is about to be covered up.

Clause 56(4) deals with work carried out on a daywork basis. Where this is the case, the contractor is paid under the conditions and at the rates and prices set out in either:

1 the daywork schedule included in the contract or, if there is no such schedule,
2 the 'Schedule of Dayworks carried out incidental to Contract Works' issued by The Federation of Civil Engineering Contractors.

In both cases, the rates current at the date of the execution of the daywork should be used.

The contractor is required to:

1 provide the engineer with records, receipts and documentation to prove the amounts paid and/or the costs incurred, and
2 submit quotations to the engineer for his approval before ordering materials, if so required.

The form and delivery times of the returns are to be as directed by the engineer, and are to be agreed within a 'reasonable time'. It is debatable as to what length of time is 'reasonable', and is hence likely to vary from contract to contract.

Clause 57 states that the '*Civil Engineering Standard Method of Measurement 1985*' will be deemed to have been used in the preparation and measurement of the bill of quantities, unless the contract documents show to the contrary. Should any amendment or modification (or different method of measurement) be adopted, it must be stated in the appendix to the form of tender.

Valuation of subcontractors' work

Subcontractors fall into two categories, nominated and domestic, as described in

Chapter 3. As a consequence, the manner of valuing their work will vary depending upon the category into which a particular subcontractor falls.

Nominated subcontractors' work is catered for in the bill of quantities by either a provisional sum or a prime cost (PC) item. In both cases a sum expected to cover the value of the nominated subcontract is inserted in the '£' column when the 'bill' is prepared. Items for work to be executed by nominated subcontractors should be followed by separate items (to be priced by the tenderer) for:

1. Labours in connection therewith, in the form of lump sums. These are to cover the main contractor's costs associated with the subcontractor's use of various site facilities or, where the subcontractor's work is to be carried out off-site, for handling and dealing with his materials as necessary.
2. All other charges and profit, in the form of a percentage.

The final amount that the main contractor eventually receives in respect of this latter item depends on the total of the subcontractor's final account since the quoted percentage is applied to that total.

The procedure adopted for the payment of nominated subcontractors is generally as follows:

1. The nominated subcontractor renders an invoice for the work carried out to the main contractor, who forwards it to the engineer (or his quantity surveyor) for checking and agreement. This is frequently done when the contractor and the engineer meet to prepare the draft interim valuation.
2. The agreed amount is then included in the next interim valuation or final account, and paid to the main contractor as part of the subsequent certificate.
3. Following receipt of the amount as part of his payment, the main contractor pays the nominated subcontractor.

Payment of nominated subcontractors is covered by Clause 59(5), (6) and (7) of the *ICE Conditions of Contract*.

Clause 59(5) states that the contract price shall include:

1. The actual price paid or due to be paid to the nominated subcontractor net of all trade and other discounts, rebates and allowances, other than discounts for prompt payment.
2. The sum (if any) provided in the bill of quantities for labours in connection therewith.
3. The percentage (previously inserted by the contractor) of the actual price paid in respect of all other charges and profit. If no such provision has been made, the rate inserted by the main contractor in the appendix to the form of tender as the percentage for adjustment of sums set against prime cost items is to be used.

Clause 59(6) requires the contractor to produce, when required by the engineer, all relevant documentation justifying expenditure in respect of nominated subcontract work.

Clause 59(7) deals with the engineer's right to demand from the main contractor before he issues any certificate, proof that previously certified sums received in respect of nominated subcontractors have been paid to those subcontractors.

If the main contractor has not paid such sums he must:

1 Give details of the reasons in writing to the engineer.
2 Produce proof that he has so informed the nominated subcontractor in writing.

If the main contractor has not paid a nominated subcontractor amounts previously certified and paid to him, the employer is entitled to pay that subcontractor direct, and to deduct, by way of set-off, such amounts from future payments due to the main contractor.

Where the main contractor employs domestic subcontractors the payment is much simpler since the agreement does not directly involve the engineer or the employer. Work carried out by this type of subcontractor is covered by the normal items and rates in the bill of quantities, and can be carried out either as a complete service whereby the subcontractor provides the labour and supplies materials and specialist plant as required, or simply does the work as a 'labour only' subcontractor. Either way, the main contractor will agree with the subcontractor the amount of work done, and pay the subcontractor without reference to the engineer.

The method of evaluation and payment will vary, and may be by daywork, lump sum or measurement, depending upon the nature of the work. Complications sometimes arise with the payment of this type of subcontractor, particularly regarding plant and services provided by the main contractor. As a result, counter charges are made and amounts deducted from payments due to the subcontractor by way of set-off.

Where subcontract work is based on a formal contract, the form commonly used is the form of subcontract provided by the Federation of Civil Engineering Contractors. This is frequently referred to as the 'blue book' due to its colour.

Variations to the contract

Variations to an existing contract fall into two categories:

1 Variations to the terms and conditions.
2 Variations to the quality and quantity of the works.

Variations to the terms and conditions may take the form of alterations made by agreement between the parties to the contract (the employer and the contractor). If the variations are extensive it may be advisable to rescind that contract and replace it with a new one. The new contract must comply with the normal rules for contracts described in Chapter 1, which include the need for consideration to pass between the parties. If only one party should benefit from the contract variations that party would:

1 have to forego some advantage under the contract, or
2 give some new service, or
3 pay a sum of money for the benefit of the other party.

Clause 2(1)(c) clearly indicates that the engineer has no powers to alter the terms and conditions of the contract, unless the employer specifically appoints the engineer to negotiate and conclude such variations on his behalf. Since the engineer is not a party to the contract, the issue of such variations would be exceeding his powers, and he could be liable to the employer for those actions and for the associated costs.

Variations to the quality and quantity of the works are common features of construction contracts and most of the standard forms of conditions of contract recognize and make provision for this by including 'variation clauses'. Such clauses empower the engineer to amend the specification and extent of the works, and bind the contractor to carry them out. Some forms limit the total value of variations that the engineer can make to a specific amount, to protect the contractor's interests. Variations beyond the limit can only be made with the express agreement of the contractor. The *ICE Conditions of Contract* do not limit the value of variations, but amendments should not be introduced that substantially change the character or scope of the contract.

Under the *ICE Conditions of Contract* variations in the quality and quantity of the works are covered by Clauses 51 and 52.

Clause 51 requires the engineer to order variations to any part of the works that is, in his opinion, necessary for the completion of the works. The engineer also has power to order any variation that for any other reason, in his opinion, is desirable for the satisfactory completion and/or improved functioning of the works. Such variations may be ordered not only prior to the issue of the substantial completion certificate, but also during the defects correction period, and may include:

1 Additions
2 Omissions
3 Substitutions
4 Alterations
5 Changes in quality, form, character or kind
6 Changes in position, dimension, level or line
7 Changes in any specified sequence, method or timing of construction

All variations to the work must be ordered by the engineer. They must also be in writing, unless the engineer considers it necessary to give an order orally, for instance, to avoid delay. Such an oral order must be confirmed in writing as soon as possible (see Chapter 4).

Clause 51 also states that no ordered variation shall vitiate or invalidate the contract and that the value of variations, unless due to the contractor's default, must be taken into account in ascertaining the contract price (final account). It concludes by stating that an order in writing is not required for changes in quantity of work unless such changes are the result of an ordered variation.

Engineers' variation orders to contractors may be in the form of 'site instructions' or 'variation orders'. Sometimes several relatively simple site instructions may be listed on a single variation order. The anticipated method of payment may well be indicated on the order form, an example of which is shown in Figure 7.1.

Clause 52 deals with the valuation of ordered variations, daywork and notice of claims. The engineer must consult with the contractor before ascertaining the value of ordered variations, which must be arrived at as follows:

1 Where work is similar to, and executed under similar conditions to work priced in the bill of quantities, at such rates and prices in the bill of quantities as may be applicable.
2 Where work is not of a similar character or is not executed under similar conditions or ordered during the defects correction period, rates and prices in the bill of quantities must be used as the basis for valuation as far as may be reasonable, failing which a fair valuation must be made.

If the engineer and the contractor fail to agree any rate or price, the engineer is required to determine a rate or price in accordance with the above principles, and to notify the contractor accordingly.

If a variation ordered under Clause 51 renders contract rates or prices unreasonable or inapplicable in the opinion of either the engineer or the contractor then that party must give notice to the other that such rates or prices should be varied before the varied work is started, or as soon as possible thereafter. The engineer will then fix rates or prices that he thinks reasonable and proper.

Additional or substituted work may be carried out on a daywork basis if the engineer considers it necessary or desirable (Clause 52(3)). The decision is the engineer's alone, and the order must be in writing. The basis of payment to the contractor is set out in Clause 56(4) and has been described earlier in this chapter. Figure 7.2 shows a completed example of such a sheet for the work covered by the variation order illustrated in Figure 7.1.

Clause 52(4) deals with notice of claims, that is, the contractor giving notice of his intention to claim a higher rate or price, or additional payment. Sometimes, the measurement of additional quantities is referred to as the measurement of a claim. This is not strictly correct, since measurement of additional quantities is provided for under the contract. A claim is more accurately described as a request for extra payment to cover costs incurred by the contractor which are additional to those that could have been reasonably foreseen and provided for under the contract. Costs include overheads and other valid charges, whether on or off the site, but do not include profit.

Claims should be presented, together with adequate substantiation, as soon as

SERIAL No. Order No. 12

VARIATION ORDER

Contract: Blackstone By-Pass Date: 3rd May 19--

Contractor: Civil Engineering Contractors Ltd.

In accordance with Clause No. 51 of the ICE Conditions of Contract you are hereby authorized to carry out/~~omit~~ the following work under the above Contract.

Remedial works to nearside kerbing at metreage 700-north bound carriageway-as directed on site.

.... Engineer/Resident Engineer

ESTIMATED COST/~~SAVING~~	Method of Payment (delete items not applicable):—
£400	1. ~~Measurement~~ 2. Daywork 3. ~~Agreed Lump Sum~~ 4. ~~Alternative to items 1-3 as stated below~~:—
ACTUAL COST/SAVING	

Figure 7.1 *Example of a variation order*

all the necessary information is available, and should state the contractual reasons on which they are based, include statements of events giving rise to the claims, and detailed calculations of the amounts. Claims have to be considered by the engineer from two standpoints: principle and quantification, in that order as there is no point spending time ascertaining an amount if there is no contractual basis for payment of that amount.

Guidance on submission and consideration of claims can be found by reference to *Notes for Guidance on the submission and consideration of Contractual Claims* produced jointly by the Department of Transport and the Federation of Civil Engineering Contractors.

Under Clause 52(4), if the contractor intends to claim a higher rate or price than that fixed by the engineer following the valuation of an ordered variation or changes in quantities in accordance with Clause 56(2), he must do so in writing within twenty-eight days of being notified of that rate or price. If, on the other hand, the contractor requires any additional payment other than as just described, he must give written notice as soon as is reasonably possible or within twenty-eight days after the events giving rise to the claim, and must keep all records necessary to support such a claim.

Upon receipt of the notice the engineer may, without admitting the employer's liability, instruct the contractor to keep such records as he may consider necessary and material to the claim, and the contractor must permit the engineer to inspect such records, and provide him with copies if instructed.

The contractor must then send to the engineer, as soon as is reasonable in the circumstances, a first interim account giving full and detailed particulars of the amount claimed at that date, and of the grounds on which the claim is based, and when required by the engineer, provide further up-to-date accounts and details. If the contractor does not comply with the above requirements, or substantially prejudices the engineer's investigation of the claim, he is only entitled to that amount which the engineer can substantiate from the information available. Any such amount must be included in any interim payment certified by the engineer.

It should be noted that many claims are the result of the contractor encountering adverse physical conditions or artificial obstructions, which the *ICE Conditions* cater for under Clause 12 (see Chapter 13).

Subclause (1) of that clause states that the contractor has to notify the engineer of any such conditions encountered, and Subclause (2) of his intention to make a claim for additional payment or extension of time 'as early as practicable'. Subclause (2) also provides an alternative whereby the notification of intention to claim can be 'as soon thereafter as may be reasonable', which appears to conflict with the twenty-eight days required under Clause 52(4)(b). A problem is unlikely to arise and the Clause 12 provision prevail over 52(4)(b), if notification of the conditions encountered has been given in accordance with Clause 12(1).

CONTRACT: BLACKSTONE BY-PASS

DESCRIPTION OF WORK: Remedial work to nearside kerbing - north bound carriageway - metreage 700 as V.O. N.°12.

DAYWORK SHEET No. 34
WEEK ENDING 6th May 19--
SIGNED BY (R.E.)
SIGNED BY AGENT S.W.

LABOUR	MON	TUES	WED	THUR	FRI	SAT	SUN	NPOT	TOTAL	RATE	£	p
Ganger - Hooker				5	3½				8½	4.00	34	00
Labourer - Allen				3	3				6	2.88	17	28
Labourer - Cook				2	1½				3½	2.88	10	08
Hymac Opr. - Shears				1½	1				2½	3.47	8	68
J.C.B. Opr. - Sweet (£2.88 plus 59p)				1	1				2	3.47	6	94
FLAT HOURS USED TO ASSESS SUBSISTENCE, BONUS & WELFARE									TOTAL 22½ HRS.		76	98
	TOTAL CARRIED TO LABOUR SUMMARY £											

PLANT & TRANSPORT	MON	TUES	WED	THUR	FRI	SAT	SUN	TOTAL	RATE	£	p
Hymac 370 (10(37))				1½	1			2½	15.36	38	40
J.C.B. 3C (10(37))				1	1			2	15.36	30	72
								TOTAL		69	12
								5 PLUS %		3	46
TOTAL FOR PLANT CARRIED TO SUMMARY £										72	58

MATERIALS	QTY.	UNIT	RATE	£	p
Concrete C 7.5P	2.65	m³	50.00	132	50
TOTAL				132	50
12½ PLUS %				16	56
TOTAL FOR MATERIALS CARRIED TO SUMMARY £				149	06

LABOUR SUMMARY	£	p
LABOUR FROM DETAIL	76	98
ADD 148 %	113	92
WELFARE HRS. at		
SUBSIST – x		
ADD + %		
TOTAL FOR LABOUR CARRIED TO SUMMARY £	190	90

SUMMARY	£	p
LABOUR	190	90
PLANT	72	58
MATERIALS	149	06
TOTAL VALUE £	412	54

Figure 7.2 *Example of a priced statement for a daywork operation*

Fluctuating price contracts

The majority of civil engineering contracts include priced bills of quantities where the rates are fixed. If the actual quantities of work carried out remain unchanged from those originally estimated and there are no ordered variations or claims the contractor is eventually paid the tender figure (the 'tender total'). However, it is usual for the quantities to change and for variations to occur, hence these contracts are remeasured as the work proceeds and valued at the rates inserted in the bill of quantities. As a consequence, they are often referred to as 'measure and value' contracts. Additionally, they may also be known as 'fixed price contracts'.

Under normal circumstances, the contractor takes all the risks of increased costs due to inflation and higher taxation, etc. Naturally, the contractor will make some provision for such increases in his tender, but difficulties arise where contracts are carried out over a long period. Pricing becomes a gamble and the tenderer who anticipates the least increases often wins the contract only to be caught out when there is a sudden upsurge in prices, sometimes with disastrous consequences.

The traditional way around this problem was for the contractor, at the time of tender, to provide a list of the principal materials to be used in the contract and the prices for those materials on which the bill rates were based. Rates for the various items of plant and labour likely to be employed would also be included. Throughout the work the contractor would keep records of the quantities of the various materials used together with the prices paid. Likewise, records would be kept of all plant used and labour employed. At the end of the contract, the contractor would be reimbursed by the employer all agreed amounts in excess of the original.

Invariably, the whole procedure was complicated, very time-consuming and involved the engineer, or his quantity surveyor, in considerable tedious and sometimes acrimonious arguments, particularly regarding the time of the increases relative to the works requirement. Into the bargain, it was a procedure widely open to misuse.

In the early 1970s alternative methods were introduced using various formulae. These are now widely used, do not require record keeping and are relatively easy to operate and agree. The formulae commonly used in building and civil engineering works were originated by the National Economic Development Office (NEDO) and indices are compiled monthly by the Department of the Environment, and published by Her Majesty's Stationery Office. They are known as the *Osborne Formula* and the *Baxter Formula*. The former is the more complicated of the two having forty-eight basic work categories and five further categories for specialist engineering installations (electrical, heating and ventilating, lifts, structural steel and catering equipment). For contracts involving mechanical and electrical plant, a third formula is used, referred to as BEAMA after its publishers, the British Electrical and Allied Manufacturers Association. It has only two elements and uses indices published by the Trade and Industry Journal.

Baxter formula

Generally, the Baxter Formula is used in civil engineering contracts that are expected to be in excess of twenty-four months (previously, twelve months) duration. Such contracts include clauses variously referred to as 'contract price adjustment', 'contract price fluctuations' or 'variation of price' (VOP). The incorporation of the clauses and the use of the Baxter Formula does not adjust, fluctuate or vary the bill rates – only the contract price (the total cost of the contract). The individual rates, therefore, remain unchanged.

It should be noted that adjustments, fluctuations and variations of prices can result in either increased or decreased costs resulting from changes in costs incurred by the contractor. It is wrong, therefore, to refer to fluctuations as 'increased costs', although in reality this is what they usually are. Furthermore, the contract price is automatically adjusted following any variation in the indices: the contractor is not required to prove additional costs nor is the engineer required to seek price reductions.

The procedure for calculating the contract price fluctuations is laid down by the issuing bodies of the *ICE Conditions of Contract*, and is broadly as follows:

1 At the time the contract is being prepared (or at the time of tender by agreement with the contractor(s)) the engineer is required to estimate the proportions of:

   (a) Labour and supervision
   (b) Plant and transport
   (c) Materials

   included in the works. The engineer is further required to break down the material element into proportions relative to twelve basic components as applicable. The total must represent 90 per cent of the works; the remaining 10 per cent representing all other costs which are not subject to any adjustment. These proportions are included in the contract documents, effectively being agreed by the parties as correct, even though, in fact, they may not be so. Figure 7.3 shows paragraph (4) of the contract price fluctuations clause indicating the various proportions applicable to a particular contract.

2 Shortly after the contract is accepted 'base index figures' are ascertained by reference to indices applicable to the various elements forty-two days prior to the date of tender.

3 Following every monthly valuation the current indices (which may be either provisional or final) are recorded, taking those applicable forty-two days prior to the date of the valuation. The 'price fluctuation factor' (PFF) is then found for each element (or proportion). All the PFFs are added together to arrive at a single total for the contract for that particular month. The following formula is used to arrive at the PFF:

$$\text{PFF} = A \times \frac{(C-B)}{B}$$

where A = contract proportion
B = base index
C = final index

Figure 7.4 shows the calculation of the price fluctuation factor for a particular month using the proportions shown in Figure 7.3.

4 Following calculation of the total PFF, the 'effective value' has to be ascertained. This is the value of the monthly valuation (before deducting retention) less amounts for:
   (a) Dayworks
   (b) Nominated subcontractors
   (c) Items based on actual cost or current prices, for example local authority business rates, resident engineer's telephone charges, etc.
   (d) Any previous fluctuations
5 Once the effective value has been established, it is multiplied by the total PFF to arrive at the price fluctuation for the month, as shown in Figure 7.4. Figure 7.5 shows the build up of the total price fluctuation over the full period of a contract.

It is worth noting that the date of issue of the certificate of completion of the works may have an important bearing on the total value of the fluctuations, since it halts the process. Work carried out subsequently will only have the indices applied that were applicable forty-two days prior to that date (see Figure 7.5). As a result, contractors often ask for completion to be effective from around the middle of the month rather than during the first or second week. By so doing they gain one month's indices increases. Conversely, during periods of falling costs, completion in the first two weeks of the month would result in indices two months earlier being applicable, which would maximize the contractor's return.

Finally, attention is drawn briefly to a special modified version of the Baxter Formula applicable to fabricated structural steelwork. The principal differences from the full formula are that it uses only the four following indices. The current index figure to be applied in each case is indicated alongside:

1 Labour employed in fabrication – fifty-six days prior to the last day of the valuation period.
2 Labour employed in erection – fourteen days prior to the last day of the valuation period.
3 Materials specifically purchased for inclusion in the works – the date of delivery to the fabricator's premises.
4 Materials not specifically purchased for inclusion in the works – the date of the last of deliveries referred to in 3 above.

In other respects the formula for fabricated structural steelwork operates in a similar manner to the full Baxter Formula.

For the purpose of calculating the Price Fluctuation Factor the proportions referred to in sub-clause (3) of this clause shall (irrespective of the actual constituents of the work) be as follows and the total of such proportions shall amount to unity:-

(a) 0.16 in respect of labour and supervision costs subject to adjustment by reference to the Index referred to in sub-clause (1)(a) of this Clause;

(b) 0.16 in respect of costs of provision and use of all civil engineering plant, road vehicles etc. which shall be subject to adjustment by reference to the Index referred to in sub-clause (1)(b) of this Clause;

(c) the following proportions in respect of the materials named which shall be subject to adjustment by reference to the relevant indices referred to in sub-clause (1)(c) of this Clause:-

0.10 in respect of Aggregates

0.03 in respect of Bricks and Clay Products generally

0.03 in respect of Cements

0.01 in respect of Cast Iron products

0.35 in respect of Coated Roadstone for road pavements and bituminous products generally

0.04 in respect of Fuel for plant to which the Gas Oil Index will be applied

0.01 in respect of Timber generally

0.01 in respect of Reinforcing steel (cut bent and delivered) and any other metal sections

Nil in respect of Fabricated Structural Steel;

(d) 0.10 in respect of all other costs which shall not be subject to any adjustment;

Total 1.00

Figure 7.3 *Paragraph (4) of the contract price fluctuation clause. It should be noted that only nine material elements are shown, and that the complete paragraph (4) also includes Fuel – DERV, Metal Sections, and Labour and Supervision in fabricating and erecting steelwork*

CONTRACT ....

CONTRACT PRICE FLUCTUATION FOR ....

VALUATION NO.............................. DATE ....

DATE OF BASE INDEX .... DATE OF APPLICABLE INDICES ....

CONSTITUENTS OF WORK	CONTRACT PROPORTION (A)	BASE INDEX (B)	CURRENT INDEX (C)	PRICE FLUCTUATION FACTOR	
				PROVISIONAL (PFF)	FINAL (PFF)
Labour	0.16	294.3	340.7		0.025226
Plant	0.16	323.3	373.7		0.024943
Aggregates	0.10	383.8	507.7		0.032282
Bricks	0.03	395.5	556.5		0.012212
Cement	0.03	373.8	537.7		0.013154
Cast Iron	0.01	393.8	491.0		0.002468
Coated Roadstone	0.35	419.7	625.4		0.171539
Fuel – Gas Oil	0.04	487.1	787.1		0.024636
Timber	0.01	376.7	462.4		0.002275
Reinforcing Steel	0.01	125.6	152.6		0.002150
Structural Steel	Nil	369.1	398.7		-
	Total Price Fluctuation Factor:				0.310885

EFFECTIVE VALUE

	£	£
Gross value of work		474,536.00
Less (a) Dayworks	97.75	
(b) Nominated sub-contractors	372.60	
(c) Items based at cost, etc.............	527.07	
(d) Previous effective value	273,119.14	274,116.56
EFFECTIVE VALUE = £		200,419.44

CONTRACT PRICE FLUCTUATION

£ 200,419.44 (effective value) x 0.3109 (PFF) = £ 62,310.40

Increase/~~Decrease~~*

*Delete that which does not apply.

Figure 7.4 *Calculation of the price fluctuation factor and the total contract price fluctuation*

CONTRACT ....

CONTRACT PRICE FLUCTUATION FOR ....

1 Certificate No	2 End of Period Date	3 Gross Value of Certificate £	4 Value of Items included at current prices etc. £	5 Previous effective value £	6 Current effective value £	7 Final PFF	8 Final amount of price fluctuation £
1	13.7.19..	32,294.35	Nil	Nil	32,294.35	0.0711	2,296.13
2	31.8.19..	73,482.14	Nil	32,294.35	41,187.79	0.1158	4,769.55
3	29.9.19..	134,164.50	Nil	73,482.14	60,682.36	0.1547	9,387.56
4	27.10.19..	162,094.15	Nil	134,164.50	27,929.65	0.1584	4,424.06
5	26.1.19..	248,577.03	Nil	162,094.15	86,482.88	0.1834	15,860.96
6	26.2.19..	217,790.12	118.47 Tel 408.60 Rates 372.60 N.S/C £899.67	248,577.03	22,313,42	0.1985	4,429.21
7	29.3.19..	274,018.81	899.67	270,890.45	2,228.69	0.2560	570.54
8	28.6.19..	474,536.00	118.47 Tel 408.60 Rates 372.60 N.S/C 97.75 D.Works £997.42	273,119,14	200,419.44	0.3109	62,310.40
Final Account	-	477,903.64	327.76 Tel 408.60 Rates 372.60 N.S/C 97.75 D.Works £1,206.71	473,538.58	3,158.35	0.3109	981.93
NB Substantial completion date 4th July		2,148,860.74	4,003.47	1,668,160.34	476,696.93	£	105,030.34

Figure 7.5 *Contract price fluctuation summary sheet for a particular project*

8 Certificates and payment

Reference ICE Clauses: 12, 47, 49, 60 and 61

Certification, and payment for the work carried out under the contract is the next step in the contractual process, after measurement and valuation as described in Chapter 7.

Procedure leading to certification

Civil engineering contracts include provision for interim valuation, certification and payment while the work is in progress. Such payments are sometimes referred to as 'on account payments'.

Normally, evaluation of the work is programmed on a monthly basis in accordance with Clause 60, which specifically requires the contractor to submit to the engineer statements at monthly intervals, showing the following:

1 The estimated contract value of the permanent works (including dayworks) up to the end of that month.
2 A list and value of goods or materials on site that are not yet included in 1 above.
3 A list and value of goods or materials identified in the appendix to the form of tender, but not yet delivered to the site. These are vested goods or materials which have become the property of the employer.
4 Estimated amounts to which the contractor considers himself entitled for which provision is made under the contract, for example claims (covered by a total of nineteen clauses in the contract) and amounts allowed under the variation of price clause, if applicable.

Also included should be amounts for nominated subcontractors. These are required to be listed separately.

The following points should be noted:

1 The contractor is not obliged to submit a statement if he does not require payment at that particular time.

2 Most contracts have a 'minimum amount of interim certificates' included in the appendix to the form of tender. If this figure is not reached during the month, the contractor is not entitled to payment until the next month, or whenever that figure is reached. The value of the work not paid for will be added to the value of the next month's work. The minimum certificate restriction does not apply during the defects correction period – payments, however small, can be made.
3 Every interim valuation includes all the work carried out to date and, therefore, provides the total value of the contract work executed up to the end of that particular period.
4 Goods and materials included in the previous 2 and 3 (page 88), that is, materials on site and vested materials off site, are not subject to retention (see page 93) although in the case of 2, a maximum percentage of the value allowed to be paid (usually 97 per cent) will be stated in the appendix.

After the engineer has received the statement, it will be checked for accuracy and content, and amendments will be made as the engineer considers necessary. The engineer will then issue a certificate for the amount 'which in the opinion of the engineer' is due. It should be noted that a certificate is considered to be the expression, in a definite form, of the exercise of judgement, opinion or skill by the engineer regarding some matter provided for within the terms of the contract. In this instance, it is the value of the work carried out.

The layout of a typical summary sheet of an interim valuation is shown in Figure 8.1. The amount for payment, or amended amount as the case may be, is the figure that the engineer will certify.

Interim and final certificates

The amount certified by the engineer will be the amount ascertained as in the previous paragraphs, less retention and any amounts previously paid to the contractor on account of that particular contract.

The complexity of evaluating contract works results in interim certificates being based, for the most part, on approximate estimates of work done. Nevertheless, the issued certificate is legally binding on the employer and must normally be paid in full. Where this is not the case, the employer must notify the contractor forthwith and give full details of how the amount being paid has been calculated.

The final certificate authorizes payment of the final account, which is the last evaluation and is, in effect, the last interim valuation. The final account must be produced by the contractor within three months of the date of the defects correction certificate (or some fifteen months after the certificate of substantial completion). Together with the final account, the contractor must provide all the necessary documentation for its verification. The engineer has a further three

NAME OF EMPLOYER...

ADDRESS OF EMPLOYER..

....

CONTRACT TITLE...

INTERIM VALUATION No......................for work up to and including

*..

Section No.	1	Preliminaries	32,670-00
	2	Site Clearance	16,790-00
	3	Fencing	9,430-00
	4	Drainage and Service Ducts	3,116-00
	5	Earthworks	150,500-00
	6	Bridgeworks	89,900-00
	7	Retaining Walls	6,205-00
	8	Testing Materials and Workmanship	716-00
			309,327-00
Dayworks			3,358-00
			312,685-00
Less Retention: 5%			15,634-25
			297,050-75
Add Materials on Site: 97% of £16,500-00			16,005-00
			313,055-75
Less Previous payments			252,713-25
		AMOUNT NOW DUE FOR PAYMENT	£60,342-50

* The date at the end of the monthly statement period

Figure 8.1 *A typical summary sheet of an interim valuation – the above interim valuation relates to a contract prepared in accordance with the* Method of Measurement for Highway Works (July 1987). *A similar valuation for a contract prepared in accordance with the* Civil Engineering Standard Method of Measurement *may be different in detail, but the basic format is the same*

months in which to check the final account and to certify the amount 'which in his opinion is finally due under the contract up to the date of the defects correction certificate.' The final account must be conclusive and not subject to later additions – it must, in fact, be final. Copies of every certificate issued by the engineer must be sent to both the employer and the contractor with, in the case of the contractor, detailed explanation as necessary.

Payments to the contractor must be within:

1 Twenty-eight days of the date on which the engineer received the interim valuation.

   This short period places considerable pressure on the engineer's staff, but is relieved to some extent where the practice of approximating two certificates out of three and of preparing the third accurately is adopted.

2 Twenty-eight days of the date on which the engineer issues the final certificate.

In the case of overpayment, the contractor must repay the employer any amounts due within twenty-eight days of the date on which the engineer issues the final certificate.

If the engineer fails to certify, or the employer fails to pay the contractor (or vice versa) within the twenty-eight day period, the employer is required to pay the contractor (or vice versa) monthly compounded interest for each day on the amount due. The rate of interest is the base rate of the bank specified in the appendix plus 2% per annum. The interest provision also applies to amounts considered by an arbitrator to have been due by a given date.

The time restrictions imposed may appear onerous at first sight, but consideration has been given, in the case of monthly payments, to easing contractors' cash flow problems and maintaining financial stability. In the case of production of the final account, three months is considered a sufficient period since it is fifteen months after issue of the certificate of completion of the works, which is something like three and a quarter years from commencement of the works for the average contract. Furthermore, admeasurement of the permanent work is normally begun soon after commencement of the work on site and so the final account is being prepared long before the issue of the certificate of substantial completion, let alone the defects correction certificate.

An example of the format of a typical engineer's certificate, authorizing payment to the contractor, is shown in Figure 8.2.

It should be noted that the engineer has the power to deduct from any certificate the value of any work previously certified, if, subsequently to that certification, he has good reason to be dissatisfied with that work.

However, the engineer cannot delete or reduce in any interim certificate amounts previously certified in respect of nominated subcontractors if the main contractor has already paid, or is bound to pay, those sums to those nominated subcontractors. Furthermore, if the engineer deletes or reduces in the final account amounts previously certified in respect of nominated subcontractors and

ENGINEER'S CERTIFICATE

Contract .... Ref: ....

Valuation No. .... End of Valuation Period .... Contractor ....

Date of Receipt of Contractor's Monthly Statement .... Address to which payment is to be sent ....

Tender Total £ ....

This Certificate must be paid by .... Financial Code ....
Interest is payable on overdue amounts (See Clause 60(7) of ICE Conditions of Contract

	£
Value of Works	
Claims	
Balancing Item ADD/DEDUCT	
Contract Price Fluctuations	
Materials on Site 97% of £..................	
Materials not on Site but vested in Employer 97% of £..................	
Valuation	
Retention	
Total Value now due	
Deduct previous certified	
AMOUNT OF THIS CERTIFICATE (EXCLUSIVE OF V.A.T.)	

Approved .... Quantity Surveyor Date ....

Recommended .... Engineer's Representative Date ....

Certified .... for Engineer Date ....

Amounts included in this Certificate in respect of Nominated Sub-Contractors (Exclusive of V.A.T.)

NAME	AMOUNT £	INCLUDED IN PREVIOUS CERT. £	INCREASE IN THIS CERT. £

FOR INFORMATION OF THE CONTRACTOR Adjustments to Engineer's Certificate by Employer

			£
	Amount Certified by Engineer		
VALUE ADDED TAX	VAT Payable on £.................. @	%	
ADD	Interest on Overdue Payments (Cl. 60(7))		
	Other matters (to be listed separately)		
	ADD		
DEDUCT	Liquidated Damages (Cl. 47)		
	Other matters (to be listed separately)		
	DEDUCT		
	PAYMENT TO CONTRACTOR		

Distribution:
- ☐ Finance (Employer)
- ☐ Contractor
- ☐ Engineer's Representative
- ☐ Project Engineer
- ☐ Q.S. - Site
- ☐ Q.S.
- ☐ Computer Group
- ☐ File

Figure 8.2 *An example of an engineer's certificate. When completed, it would authorize payment to the contractor*

the main contractor has already paid those sums, the employer must reimburse the main contractor the amounts overpaid.

The points mentioned in the last paragraph have tended to discourage many engineers from using nominated subcontractors and PC items in contracts in recent years – a practice which is generally approved by main contractors.

Retention

Retention is money deducted from the amounts due to contractors. It is retained by the employer for a stated period to provide some form of protection against incorrect or substandard workmanship and materials. At the end of the stated period it is paid to the contractor provided that any faults discovered are corrected or substandard materials are replaced.

Retention is expressed as a percentage of the work carried out and is calculated on, and deducted from, every interim valuation. The applicable percentage is indicated in the appendix, and the total amount of retention calculated must not exceed the limit also set out therein. It is recommended that the percentage should not be in excess of 5%, and the limit should not exceed 3% of the tender total.

The following example indicates how retention is calculated and the limit may be applied.

A contract with a tender total of £200,000, 5% retention, and a 3% limit.
The maximum retention will be £6,000.

£6,000 at 5 per cent will be reached when contract work to the value of £120,000 has been carried out. Therefore, prior to the executed work reaching £120,000, 5 per cent retention will be deducted, but from then on, retention will not exceed £6,000.

Upon the issue of a certificate of substantial completion by the engineer half the retention is released. This amount must be paid to the contractor within fourteen days. The remaining half is paid to the contractor at the end of the defects correction period, following the issue of the defects correction certificate. This half of the retention, like the first half, must be paid within fourteen days of the certificate being issued, notwithstanding that there may be outstanding claims by the contractor against the employer.

Liquidated damages

Liquidated damages are damages based on a genuine pre-estimate of the costs likely to be incurred by one party (the employer) in the event of a breach of contract by the other party (the contractor). For the estimated amount to be classified as liquidated damages it must be reasonable. It will then be recoverable in full, although in reality the actual loss may differ from that anticipated. An amount in excess of that arrived at as stated above would be considered to be a 'penalty' (a punishment). As such it is not reckoned to be liquidated damages, and would not be recovered in full.

Clause 47 provides for liquidated damages to be paid by the contractor in the event of delayed completion of the works. A sum per day or per week may be included in the appendix to the form of tender as recoverable by the employer from the contractor if the contract is not completed within the contract period (or the permitted extended time for completion), assuming that responsibility for the over-run does not rest with the employer or the engineer. Further sums may also be included for sectional completion where applicable.

Since all sums are stated in the contract documents, and tenders are submitted in full knowledge of them, they are considered to be mutually agreed. Subclause (4)(a) states that the total amount of liquidated damages payable is limited to the sum stated in the appendix, and continues: 'If no such limit is stated therein liquidated damages without limit shall apply'. This latter sentence is confusing and is possibly in conflict with (4)(b) which says: 'Should there be omitted . . . any sum required to be inserted therein . . . or if any such sum is stated to be "nil" then . . . damages shall not be payable.'

The following provides an example of the calculation of liquidated damages for the whole of the works:

Time for completion for the whole of the works	100 weeks
Liquidated damages for delay	£1,000 per week
Actual time for completion	112 weeks
Extension of time granted under Clause 44	4 weeks

Amount of liquidated damages:		
Actual time		112 weeks
Less time for completion		100 weeks
Over-run		12 weeks
Less extension of time		4 weeks
	Net delay	8 weeks

∴ Liquidated damages = 8 weeks
@ £1,000 per week = £8,000

If, after liquidated damages have become payable, the engineer issues a variation order or Clause 12 conditions (adverse physical conditions, etc.) are encountered that are outside the contractor's control and such will further delay the works, the engineer must so inform both the contractor and the employer in

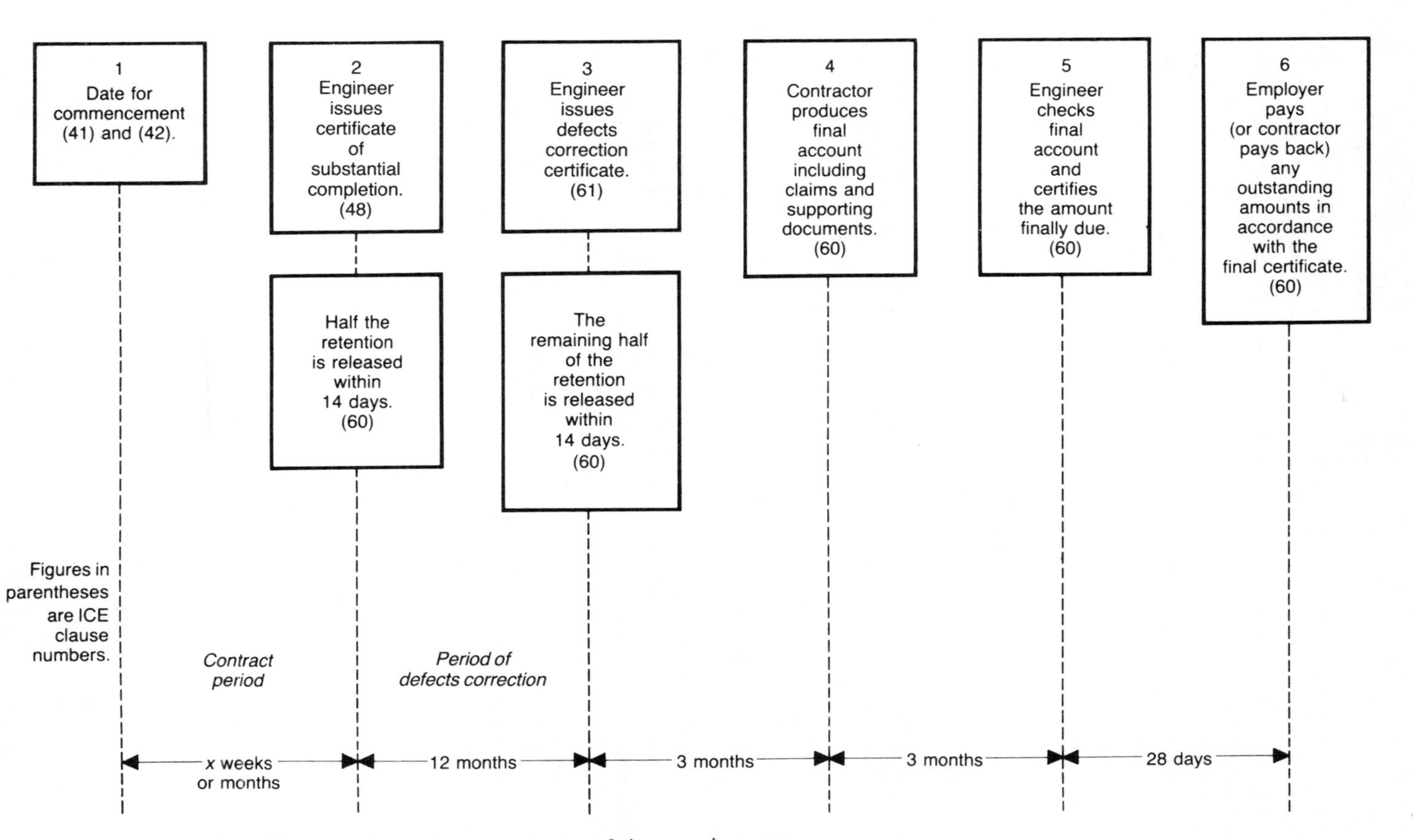

Figure 8.3 *Issue of certificates and payment – sequence and time requirements*

writing. The employer's right to liquidated damages will be suspended until the engineer's formal notification that the delay has come to an end. The employer will still be entitled to any prior accrued, and any future, liquidated damages (Subclause (6)).

The normal method of payment of liquidated damages (£8,000 in the above example) is by deduction from the final account figure, after all other adjustments have been made. Subclause (2), however, permits the employer to deduct liquidated damages at other times during the contract. Where this is the case, and if monies were wrongly deducted (owing, perhaps, to the engineer's tardy grant of an extension of time), the employer is required to reimburse the contractor the overpaid amount together with compounded monthly interest (Subclause (5)).

Defects correction certificate

The period of defects correction commences with the issue of the certificate of substantial completion and is usually of twelve months duration. At the end of the period the engineer, accompanied by the contractor, inspects the works and prepares a list of all outstanding works, together with all necessary repairs, alterations and making good, etc., that are the contractor's responsibility.

These works should be carried out within fourteen days (see Clause 49). The engineer then, in accordance with Clause 61, issues the defects correction certificate which states the date on which the contractor completed his contractual obligations to construct and complete the works to the engineer's satisfaction.

The issue of the defects correction certificate does not relieve either party from their obligations under the contract. Hidden or latent defects that later come to light mean that the contract has not been fully performed despite the issue of certificates, and contractual obligations remain enforceable throughout the period of limitation (see Chapter 1).

9 Settlement of disputes – conciliation and arbitration

Reference ICE Clauses: 66 and 67, the ICE Conciliation Procedure (1988), and the ICE Arbitration Procedure (England and Wales) (1983)

Disputes between the parties to a contract frequently occur. In the normal course of events these are satisfactorily resolved, either during the contract period or soon after completion of the contract works. Occasionally, however, disputes fail to be resolved and the parties must then seek a solution in the courts through the normal processes of the law. Alternatively to litigation, a dispute may be settled by conciliation or by arbitration as is usually the case where disputes are of a technical or commercial nature, for example in construction contracts and shipping.

Arbitration generally

Arbitration may be defined as the resolving of an issue, or issues, upon which the parties have failed to agree, by way of a solution imposed by one or more impartial persons.

The impartial person/s is/are normally chosen by the parties in dispute. That person, once appointed, is known as the arbitrator, and has powers to take evidence, call witnesses, examine documents and carry out investigations, etc., that is, all the powers necessary to allow him to come to a decision and make an award.

The ICE (and the JCT) standard form of contract provides for the settlement of disputes by a single arbitrator, whereas the FIDIC requires at least one. By way of comparison, shipping disputes employ three arbitrators.

Arbitration proceedings are independent of the courts, but fall within the normal requirements of the law: the court, in fact, has supervisory jurisdiction over the proceedings. If it is necessary, the arbitrator's award will be enforced by the courts in a like manner to a court judgement.

Some contracts, such as those incorporating the *ICE Conditions of Contract*, contain a provision that disputes will be referred to arbitration. Others do not, and a decision to arbitrate may be made only after a dispute has arisen. Alternatively, contracts may stipulate that the parties must arbitrate before litigation. Other contracts may simply provide for either arbitration or litigation as required. If it is a provision of a contract that a party should go to arbitration before seeking redress through the courts, and if one party does not go, he is technically in breach of contract. The other party is then entitled to a stay of proceedings which results in the original party being forced into arbitration.

Advantages of arbitration

Arbitration is reckoned to have several advantages over litigation. Arguably some of those advantages have been eroded over the years as disputes have become more complex and the need to win more urgent. Nevertheless, arbitration has its merits, and is still a suitable means of resolving technical disputes.

The advantages over litigation are generally reckoned to be:

1. The arbitrator is usually a technical expert, chosen by the parties who will be confident that he understands the technicalities of the problem. If the matter had to go to court the parties might not be happy that the complexities of the dispute were fully understood.
2. The use of technical experts is likely to result in earlier decisions, which in turn is likely to result in less costly proceedings than those of litigation.
3. Arbitration proceedings are arranged to suit individual disputes in a way that provides greater flexibility than court proceedings. This particularly applies to the location and length of hearings, and to the number and choice of persons involved.
4. Arbitration proceedings are more private and less formal, hence producing a more relaxed situation, with the result that hearings are conducted in a more conducive atmosphere.
5. The award of the arbitrator is final and binding unless the parties agree otherwise or an 'appeal' is made on a technicality. Consequently, disputes do not become protracted through 'appeals' as may be the case in litigation.

Despite the foregoing, arbitration may not be appropriate in some instances, particularly where disputes concern the niceties of the law, the construction of contracts, the character of parties, and where third parties are involved who are unwilling to participate in the arbitration procedure. In such cases the proceedings are likely to be held in court.

ICE requirements

If a dispute that falls outside the provisions of the *ICE Conditions of Contract* arises between the contractor and the employer relating to any matter in connection with or arising out of the contract, or carrying out the works, the matter must be referred to the engineer to settle in accordance with Clause 66.

Disputed matters include any:

Decision
Opinion
Instruction
Direction
Certificate
Valuation
} of the engineer

and may occur at any time between the commencement of the works and the ending, abandonment or breach of the contract.

There are four principal steps in the procedure as follows:

1 *Notice of Dispute*. One party serves a written 'Notice of Dispute' on the engineer stating the nature of the dispute. This can be done only if that party has taken the steps or involved the procedures available elsewhere in the contract in connection with the subject matter of the dispute. Furthermore, it cannot be done before the other party or the engineer has:
   (a) taken such steps as may be required, or
   (b) been allowed a reasonable time to take such action.
   These restrictions are intended to discourage the parties from rushing into the formal procedures before all the informal avenues have been explored.
2 *Engineer's Decision*. The engineer must settle the dispute and give his decision thereon in writing to both the employer and the contractor within:
   (a) One calendar month, where the certificate of substantial completion of the whole of the works *has not* been issued, or
   (b) Three calendar months, where the certificate of substantial completion of the whole of the works *has* been issued.
   These periods commence with the 'Notice of Dispute' being served. The 'decision' of the engineer is final and binding, unless:
   (a) the matter in dispute is referred to a conciliator and his recommendation is accepted, or
   (b) the engineer's decision is revised by an arbitrator.
3 *Conciliation*. If either party is dissatisfied with the engineer's 'decision' or the time limit for such has expired, and no 'Notice to Refer' to arbitration has been served, either party may give a written notice to the other requiring that the dispute be considered by a conciliator in accordance with the Institution of Civil Engineers Conciliation Procedure (1988) (see page 101). Subclause (5) states that the recommendation of the conciliator shall be deemed to have been accepted unless a written 'Notice to Refer' to arbitration is served within one calendar month of its receipt.
4 *Arbitration*. The dispute may be referred to arbitration, subject to the following Subclause (6) requirements and time constraints:
   (a) Where the certificate of substantial completion of the whole of the works *has not* been issued, and either party is:
      (i) dissatisfied with the engineer's 'decision', or

(ii) the time limit of one calendar month for an engineer's 'decision' has expired, or
(iii) either party is dissatisfied with the conciliator's recommendation.

The time constraints are:

(i) within *three calendar months* after receipt of the engineer's 'decision',
(ii) within *three calendar months* after expiry of the one calendar month for an engineer's 'decision', and
(iii) within *one calendar month* of the receipt of the conciliator's recommendation.

(b) Where the certificate of substantial completion of the whole of the works *has* been issued the provisions in (a) above apply, except that *one calendar month periods are changed to three calendar months*.

Following the dispute being referred to arbitration (serving of the 'Notice to Refer'), the employer and the contractor are required, jointly, to appoint a single arbitrator. If they fail to agree, and to appoint within one calendar month of one party serving a written 'Notice to Concur' on the other, either party may apply to the President of the Institution of Civil Engineers to make such an appointment.

If the chosen arbitrator declines the appointment, or if after appointment is removed by a court order, or is incapable of acting, or dies, the vacancy should be filled within one calendar month of the vacancy arising. Failure to appoint a substitute within the stated period will result in an appointment being made by the President of the Institution of Civil Engineers if either party requests him so to do. If the President is unable to exercise the above functions they will be carried out by the Vice-President.

The subsequent arbitration will be governed by the Arbitration Acts 1950–1979 or any statutory re-enactment or amendment thereof, and will be conducted in accordance with the Institution of Civil Engineers Arbitration Procedure (1983) (see page 102).

The powers of the arbitrator are extensive, and he is specifically given full power to:

1 Open up, and
2 Review and revise any of the engineer's:
   Decisions
   Opinions
   Instructions
   Directions
   Certificates
   Valuations

Neither party is limited in the arbitration proceeding to the evidence or arguments that they put before the engineer – in other words, they may put new or additional evidence before the arbitrator. The parties must fully comply with arbitrator's requirements to enable him to arrive at a decision and an award which is binding on the parties.

The following points should be noted:

1 The word 'shall' in Clause 66(1) tends to indicate that the voluntary procedure of arbitration (as opposed to litigation) becomes somewhat involuntary once the contract has been made. In fact, the parties are free to resolve any disagreements outside of arbitration, although if one party invokes Clause 66 arbitration follows.
2 The engineer's 'decision' called for under Clause 66 is final. The contractor, therefore, should be certain that an engineer's 'decision' is what he really wants at that stage in the dispute, for once given there is no room for manoeuvre or compromise unless or until the 'decision' is revised at a later date by a conciliator or an arbitrator.
3 Unless the parties agree otherwise in writing, arbitration may take place before completion of the works.
4 No previous 'decision' given by the engineer shall disqualify him from being called as a witness and giving evidence before the arbitrator.
5 The choice of conciliation or arbitration usually rests with the first party to serve a notice, and can be served with the 'Notice of Dispute' if so required.
6 Following the engineers 'decision', the parties may proceed directly to arbitration, hence by-passing the conciliation procedure.

Finally, if the works are situated in Scotland, Clause 67(2) states that the applicable Act and procedure is the Arbitration (Scotland) Act 1894, and the Institution of Civil Engineers' Arbitration Procedure (Scotland) (1983).

ICE Conciliation Procedure (1988)

Where a dispute between the employer and the contractor has arisen in connection with the works it may be settled by conciliation under Clause 66(5), in accordance with the ICE Conciliation Procedure (1988). The Procedure's key points are as follows:

1 Either party may by written notice to the other party (accompanied by a brief statement of the matters of the dispute together with the relief and remedies sought) request that the dispute be referred to a conciliator for his recommendation.
2 The parties then have fourteen days to agree upon a person to conciliate. Failing agreement, the President of the ICE may be called upon by either party to make an appointment – again within a fourteen day period.
3 A copy of the notice, together with the names and addresses of the parties' representatives, must be sent to the conciliator.
4 The conciliator is required to proceed with all despatch and use his best endeavours to conclude the conciliation as soon as possible and, in any event, within two months of his appointment.
5 Written submissions stating the parties' versions of the dispute may be sent

to the conciliator together with their views as to the rights, liabilities and financial consequences. Copies of all relevant documents, which may be accompanied by written statements of evidence, must be included with the written submissions.

6 The conciliator may, providing he gives at least twenty-four hours notice to the parties, visit and inspect the site or the subject matter of the dispute, and may acquire information in any way he thinks fit, including meeting the parties separately.
7 The conciliator may call meetings that must be attended by the parties, providing at least seven days notice of such meetings is given. The conciliator may at such meetings, take evidence and hear submissions and may, and shall if so requested, seek legal and other advice.
8 Normally, within twenty-one days of concluding meetings, the conciliator must prepare his recommendation as to the disposal and settlement of the dispute, including any sums of money which should be paid by the parties. At the same time, or within seven days of giving a recommendation, the conciliator may, if he considers it appropriate, submit in a separate document his written opinion on the matters referred to him, and the reasons for, and comments thereon as he deems appropriate. The conciliator may also at his discretion, and if he considers it appropriate, give his preliminary views.
9 When the conciliator has prepared his recommendation he will notify the parties in writing, and send them an account for his fees, which will normally be shared equally between the parties, and be paid within seven days of receipt. Provision is made for a party defaulting in payment, whereby the other party may pay the fee due and recover the amount from the defaulting party.
10 The conciliator will not be appointed as arbitrator in any subsequent arbitration between the parties on any matter arising out of the same contract, except with the parties' agreement.
11 All documents required to be sent to the parties or the conciliator must be sent by recorded delivery.

ICE Arbitration Procedure (1983)

The Institution of Civil Engineers Arbitration Procedure (1983) was prepared by that body for the settlement of civil engineering disputes by arbitration under Clause 66, but is also suitable for other engineering arbitrations. It supersedes the 1973 procedure specifically mentioned in the *Conditions of Contract* (fifth edition).

The objectives of the procedure are twofold:

1 To relieve the President of the Institution of the burden of procedural matters.
2 To provide a vehicle for hearing disputes where the parties are unable or do not want to make their own arrangements.

The procedure, comprising nine parts (A–J but excluding I) and a total of twenty-seven rules, is described below in edited form.

Part A Reference and appointment

Rule 1 Notice to refer

A dispute or difference arises when a claim or assertion made by one party is rejected by the other party and that rejection is not accepted. Either party may then invoke arbitration by serving a 'Notice to refer' on the other party.

The 'Notice to refer' must include:

1. A list of the matters which the issuing party wishes referred to arbitration.
2. The date when those matters were referred to the engineer for his decision.
3. The date when the engineer gave his decision.
4. A statement that the engineer failed to give his decision (if applicable).

Rule 2 Appointment of sole arbitrator by agreement

After the 'Notice to refer' has been served, either party may serve upon the other a 'Notice to concur' in the appointment of an arbitrator, listing the names and addresses of persons he proposes as arbitrator.

Within fourteen days the other party must:

1. Agree in writing to the appointment of one of those listed above, or
2. Propose alternative persons.

Once agreement has been reached, the issuing party must write to the selected person inviting him to accept the appointment. A copy of the 'Notice to refer' together with documentary evidence of the other party's agreement should be enclosed with the invitation.

If the selected person accepts the appointment he must notify both parties in writing. The date of posting or service of notification, as the case may be, will be deemed to be the date of the arbitrator's appointment.

Rule 3 Appointment of sole arbitrator by the President

The parties have one calendar month from service of the 'Notice to concur' to appoint an arbitrator. If they fail to do so, either party may apply to the President to appoint an arbitrator. The parties may also agree to apply to the President without a 'Notice to concur'.

The application must be in writing and must include:

1. The 'Notice to refer'.
2. The 'Notice to concur'.
3. Any other relevant documents.
4. The appropriate fee.

The Institution will send a copy of the application to the other party, stating that the President intends to make an appointment on a specified date. The President will then contact a suitable person, obtain his agreement to arbitrate, and make an appointment. The Institution will, as soon as possible thereafter, notify both parties and the arbitrator in writing.

Rule 4 Notice of further disputes or differences
At any time before the arbitrator's appointment is completed, either party may refer further disputes or differences to him by means of an additional 'Notice to refer' in accordance with Rule 1.

Once the arbitrator has been appointed, he has jurisdiction over any issue connected with, and necessary to, the determination of any dispute already referred to him.

Part B Powers of the Arbitrator

Rule 5 Power to control the proceedings
The arbitrator is given authority to exercise all of the express and implied powers set out in this procedure as he thinks fit. The terms specifically include:

1 Orders as to costs.
2 Time for compliance.
3 The consequences of non-compliance.

The powers under this procedure are additional to any other powers available to the arbitrator.

Rule 6 Power to order protective measures
The arbitrator has the power to:

1 Give directions for the detention, storage, sale or disposal of the whole or any part of the subject matter in dispute at the expense of the parties.
2 Give directions for the preservation of anything which is likely to be required as evidence.
3 Order the deposit of money, etc. to secure the whole or any part of the amount(s) in dispute.
4 Make an order for security for costs in favour of the various parties.
5 Order his own costs to be secured.

Money ordered to be paid as above must be paid immediately into a separate bank account in the name of a stakeholder appointed by and to the directions of the arbitrator.

Rule 7 Power to order concurrent hearings
Where disputes arise under more than one contract and are concerned with the same subject matter and the parties have been referred to the same arbitrator, that

arbitrator may, with the agreement of all parties, or upon the application of one of the parties (who must be a party to all the contracts), order that the relevant matters be heard together in such a manner as he thinks fit.

The awards for concurrent hearings must be made and published separately, unless the parties agree otherwise. However, the arbitrator may prepare one combined set of reasons to cover all the awards.

Rule 8 Powers at the hearing

The arbitrator may hear the parties and their supporters at any time or place, and may adjourn the arbitration for as long as he thinks necessary if requested to do so by any of the parties.

The parties may be represented by any person including, in the case of a company or legal entity, directors, officers, employees and beneficiaries.

A person shall not be prevented from representing a party because he is also a witness, and nothing shall prevent a party being represented by different persons at different times.

The arbitrator may start the arbitration at any time after his appointment is completed.

Any meeting with, or summons before the arbitrator, may be treated as part of the hearing, provided both parties are represented.

Rule 9 Power to appoint assessors or to seek outside advice

The arbitrator may appoint a wide variety of persons to assist him, and may seek advice on any matter arising out of, or in connection with, the arbitration. The arbitrator may also, or alternatively, rely on his own knowledge and expertise.

Part C Procedure before the hearing

Rule 10 The preliminary meeting

The arbitrator will call the parties to a preliminary meeting as soon as possible after accepting the appointment to instruct them about the procedure to be followed in the arbitration.

At that meeting they will consider whether, and to what extent:

1 Part F (Short procedure) or Part G (Special procedure for experts) of the rules is to apply.
2 The arbitration may proceed on documents only.
3 Progress may be facilitated and costs saved by determining some of the issues in advance of the main hearing.
4 The parties should enter into an exclusion agreement in accordance with S.3 of the Arbitration Act 1979.

They should also consider other steps that may minimize delay and expedite the determination of the real issues between the parties.

The parties may themselves agree directions and submit them to the arbitrator for his approval. In so doing, they must state whether or not they wish Parts F or G to apply. The arbitrator will then approve the directions submitted, vary them or submit his own.

Rule 11 Pleadings and discovery
The arbitrator may order the parties to deliver pleadings or statements of their cases in any form he thinks appropriate, and order them to answer the other party's case, and to give reasons for any disagreement.

The arbitrator may order the parties to deliver in advance of formal discovery, copies of any documents which relate to matters in any pleading, statement or answer.

Any of the above must contain sufficient detail for the other party to know the case he has to answer. If sufficient detail is not provided the arbitrator may demand further and better particulars.

If a party fails to comply with any order made under this rule the arbitrator has the power, following a warning to the party in default, to debar that party from relying on the matters in respect of which he is in default, and the arbitrator may make his award accordingly.

Rule 12 Procedural meeting
The arbitrator may call procedural meetings at any time, requesting particular persons to attend, to identify or clarify the issues to be decided, and the procedures to be adopted.

Either party may apply at any time to appear before the arbitrator on any interlocutory matter, and he may call a meeting for that purpose or deal with it in any other way he thinks fit.

At any procedural meeting, or otherwise, the arbitrator may give directions for the proper conduct of the arbitration. Such directions may include an order that either or both parties shall prepare in writing and shall serve upon the other party and the arbitrator any, or all, of the following:

1 A summary of that party's case.
2 A summary of that party's evidence.
3 A statement or summary of the issues between the parties.
4 A list and/or a summary of the documents relied upon.
5 A statement or summary of any other matters likely to assist the resolution of the disputes or differences between the parties.

Rule 13 Preparation for the hearing
In addition to his powers under Rules 11 and 12 the arbitrator also has power to order:

1 That the parties agree facts as facts and figures as figures where possible.
2 The parties to prepare an agreed bundle of all relevant documents which shall

be deemed to have been entered in evidence without further proof and without being read out at the hearing, on the understanding that either party may at the hearing challenge the admissibility of any documents in the agreed bundle.

3 That any experts whose reports have been exchanged before the hearing shall be examined by the arbitrator in the presence of the parties or their legal representatives. In such cases, either party may put cross-examination or re-examination questions to a party's expert after all experts have been examined by the arbitrator, provided prior notice of the nature of the question(s) has been given.

Before the hearing the arbitrator may, and must if requested to do so by the parties, read the documents to be used. In order to do this it may be necessary to deliver the documents to the arbitrator when, and at whatever place he may specify.

Rule 14 Summary awards

The arbitrator may at any time make a 'Summary award', having power to award payment by one party to another of a reasonable proportion of the final nett amount that the party is eventually likely to be ordered to pay. The amount should take into account any defence or counterclaim to which the other party may be entitled.

The arbitrator has the power to order the above payment, or part thereof, to be paid to a stakeholder, and if such an order is not complied with the arbitrator may instruct the whole sum in the 'Summary award' be paid to the other party.

The arbitrator also has the power to order payment of costs in relation to a 'Summary award'.

A 'Summary award' is final and binding on the parties unless or until varied by subsequent awards. Any subsequent award may order repayment of monies paid in accordance with the 'Summary award'.

Part D Procedure at the hearing

Rule 15 The hearing

At or before the hearing, and following representations made on behalf of each party, the arbitrator will decide the order in which the cases will be presented and issues decided. He may order any submission or speech to be put into writing and delivered both to him and to the other party. Such submission or speech may be enlarged upon orally.

Any issue may be heard or determined separately.

If a party fails to appear at a hearing, after having had notice of the hearing or the arbitrator being satisfied that all reasonable steps had been taken to notify a party, the hearing will proceed in his absence, with the arbitrator taking all reasonable steps to ensure a fair and just outcome.

Rule 16 Evidence

The arbitrator may order a party to submit an advance list of the witnesses he intends to call at the hearing. Such a list does not mean that all listed witnesses must be called. Neither does it prevent further witnesses being called.

Expert evidence is only admissible by leave of the arbitrator, and only then on his terms and conditions. Such terms (unless the arbitrator orders otherwise) shall be deemed to include a requirement that a report from each expert containing the substance of the evidence to be given shall be served upon the other party within a reasonable time before the hearing.

The arbitrator may order disclosure or exchange of proofs of evidence relating to factual issues, and may also order the parties to prepare and disclose advance lists of points or questions to be put to witnesses.

Where lists of questions are disclosed, the party making disclosure is not bound to put any of the questions to witnesses unless ordered to do so by the arbitrator. Where a party making disclosure puts an unlisted question to a witness the arbitrator may disallow the costs thereby occasioned.

The arbitrator may order that any proof of disclosed evidence shall stand as the evidence in chief of the deponent if the other party has been given an opportunity to cross-examine the deponent thereon. The deponent may be ordered, before such cross-examination, to deliver written answers to questions arising out of the proof of evidence.

The arbitrator may put questions to any witness and/or require the parties to conduct enquiries, tests or investigations. The parties may likewise ask the arbitrator to conduct or arrange for any enquiry, test or investigation.

Part E After the hearing

Rule 17 The award

After having considered all the evidence and submissions the arbitrator will prepare and publish his 'Award', and will inform the parties in writing, specifying how and where it may be taken up upon payment of his fee.

Rule 18 Reasons

The arbitrator may, at his discretion, state his reasons for all or part of his 'Award'. Such reasons may form part of or be separate from the 'Award'.

If a party asks for reasons he must state the purpose for the request. If the purpose is to use them for an appeal, he must also specify the points of law with which he wishes the reasons to deal, and the arbitrator must give the other party an opportunity to specify additional points of law to be dealt with.

Reasons prepared as a separate document may be delivered with or later than the 'Award'.

If the arbitrator decides not to state his reasons he must keep such notes as will enable him to prepare reasons later if so ordered by the High Court.

Rule 19 Appeals
The arbitrator must be notified immediately if any party applies for leave to appeal against the award or decision to the High Court.

Once any award or decision has been made and published the arbitrator is not obliged to make any statement other than in compliance with an order of the High Court under S.1(5) of the Arbitration Act 1979.

Part F Short procedure

Rule 20 Short procedure
If the parties agree, the arbitration will be conducted in accordance with this 'Short procedure'.

Each party must set out his case in the form of a file containing:

1 A statement as to the orders or awards he seeks.
2 A statement of his reasons for being entitled to such orders or awards.
3 Copies of any documents and statements on which he relies identifying the origin and date of each document.

Copies of the file must be delivered to the other party and to the arbitrator as directed by the latter.

After reading the cases the arbitrator may visit the site or works, and is entitled to ask for further documents and information in writing if required.

Within one calendar month of completing the foregoing the arbitrator must arrange a meeting with the parties for the purpose of:

1 Receiving oral submissions and/or
2 The arbitrator putting questions to the parties, etc.

The arbitrator must give notice of any person he wishes to question, but such persons are not bound to appear before him.

Within one calendar month of concluding the above meeting the arbitrator shall make and publish his 'Award'.

Rule 21 Other matters
Unless otherwise agreed by the parties, the arbitrator has no powers to award costs. The arbitrator's fees will be shared equally between them.

Either party may, at any time before the 'Award' is published, give written notice to the arbitrator and to the other party that he wishes the arbitration by short procedure to cease. Except as follows, the short procedure will not apply or bind the parties, although any evidence given will be admissible in further proceedings.

The party giving written notice, as in the previous paragraph, is liable to pay:

1 The whole of the arbitrator's fees and charges up to the date of the notice and
2 A sum to be assessed by the arbitrator as reasonable compensation for the

costs incurred by the other party up to the date of the notice.

Full payment of all such charges will be a condition precedent to that party proceeding further in the arbitration, unless the arbitrator directs otherwise, which he should not do if non-payment prevents the other party proceeding in the arbitration.

Part G Special procedure for experts

Rule 22 Special procedure for experts

Where the parties agree, the hearing and determination of issues of fact which depend upon experts' evidence will be conducted in accordance with this 'Special procedure'.

Each party must set out his case in the form of a file containing:

1 A statement of the factual findings he seeks.
2 A report or statement from, and signed by, each expert upon whom that party relies.
3 Copies of any other documents referred to in each report or statement identifying the origin and date of each document.

Copies of the file must be delivered to the other party and to the arbitrator as directed by the latter.

After reading the cases the arbitrator may visit the site or works, and is entitled to ask for further documents and information in writing if required.

The arbitrator must then fix a meeting with the experts. At that meeting each expert may address the arbitrator and question experts representing the other parties, and be given adequate opportunity to explain his opinion and to comment upon opposing opinions.

Thereafter, the arbitrator may make and publish an 'Award' setting out, with such details or particulars as is necessary, his decision.

Rule 23 Costs

The arbitrator may in his 'Award' order payment of costs relating to the foregoing matters, including his own fees and charges.

Unless otherwise agreed by the parties, neither party is entitled to any legal costs relating to the hearing and determination of factual issues by this special procedure.

Part H Interim arbitration

Rule 24 Interim arbitration

Where the arbitration is to proceed before completion of the works the following additional provisions will apply (except in the case of a dispute under Clause 63 of the *ICE Conditions of Contract*), and the arbitration will be called an 'Interim arbitration'.

The arbitrator is obliged to make his 'Award' or 'Awards' as quickly as possible so as to allow timely completion of the works.

If an 'Interim arbitration' is not completed before completion of the works, the arbitrator must within fourteen days of completion of the works make and publish his 'Award', findings or 'Interim decision', based on the evidence and submissions up to that date plus any further evidence and submissions that the arbitrator may be prepared to receive during the above fourteen day period. The fourteen day period may be waived if the parties so agree and notify the arbitrator accordingly within that period.

In an 'Interim arbitration' the arbitrator may make and publish any or all of the following:

1 A 'Final award' or 'Interim awards'.
2 Findings of fact.
3 A 'Summary award'.
4 An 'Interim decision'.

Awards and findings are final and binding upon the parties in any subsequent proceedings. Anything not expressly identified as falling under 1, 2 or 3 will be deemed to be an 'Interim decision'. Except as previously stated, an 'Interim decision' cannot be made without the arbitrator first notifying the parties of his intentions to do so.

An 'Interim decision' is final and binding upon the parties and the engineer (if any) until the works have been completed or any 'Award' or decision is given under this rule. Thereafter, the 'Interim decision' may be re-opened by another arbitrator, and where such other arbitrator is also the arbitrator appointed to conduct the 'Interim arbitration' he will not be bound by his earlier decision.

The arbitrator in an 'Interim arbitration' has the power to direct that Part F (Short procedure) and/or Part G (Special procedure for experts) shall apply to the 'Interim arbitration'.

Part J Miscellaneous

Rule 25 Definitions

The following terms are defined for the purposes of these rules:

1 Arbitrator.
2 Institution.
3 ICE Conditions of Contract.
4 Other party.
5 President.
6 Procedure.
7 Award.
8 Interim arbitration.

Rule 26 Application of the ICE procedure
The procedure will apply to the arbitration if:

1 The parties at any time so agree.
2 The President when making an appointment, so directs.
3 The arbitrator so stipulates at the time of his appointment.

Where this procedure applies by virtue of stipulation under 3 above, the parties may within fourteen days of that appointment agree otherwise. In this event, the arbitrator's appointment will be terminated, and the parties will pay his reasonable charges equally between them.

This procedure does not apply to arbitrations under the Law of Scotland for which a separate procedure is available – *ICE Arbitration Procedure (Scotland)*.

Where an arbitration is governed by the law of a country other than England and Wales this procedure will apply to the extent that the applicable law permits.

Rule 27 Exclusion of liability
This rule excludes the ICE, its servants, agents and President of liability to any party for any act, omission or misconduct in connection with any appointment made, or any arbitration conducted, under this procedure.

10 Determination of the contractor's employment

Reference ICE Clause 63

Insolvency, bankruptcy and liquidation

The requirements of the ICE conditions regarding the determination of the contractor's employment are set out under Clause 63, which makes reference to 'bankruptcy' and 'liquidation', both of which need a brief explanation.

A person having insufficient funds to pay his debts is said to be 'insolvent'; whereas the inability to pay debts is a state of 'insolvency'. Insolvency is not bankruptcy although, clearly, it may result in such. A 'bankrupt' is a person who cannot pay his debts, and who is adjudicated by a court order to be a bankrupt; whereas 'bankruptcy' is the court proceedings for the distribution of an insolvent person's property among his creditors, and to relieve him of his liabilities in relation to the remaining amounts unpaid.

Liquidation is the winding-up of a company or partnership, that is, the cessation of business operations with a view to realizing the value of the assets, discharging liabilities and distributing proceeds among the members. A business is wound-up under the Companies Act 1948, whereas a partnership is wound-up either by voluntary agreement between the partners or by a court order.

ICE requirements

Clause 63 states that the employer may, after giving the contractor seven days notice in writing, expel him from the site and the works, without avoiding the contract or releasing the contractor from any of his obligations or liabilities under the contract if the contractor defaults in any of the following ways:

1 Becomes bankrupt.
2 Has a receiving order or administration order made against him.
3 Presents his petition in bankruptcy.
4 Makes an arrangement with or assignment in favour of his creditors.
5 Agrees to carry out the contract under a committee of inspection of his creditors.

6 Goes into liquidation (other than voluntarily for amalgamation or reconstruction).
7 Assigns the contract without the written consent of the employer.
8 Has an execution levied on his goods, which is not stayed or discharged within twenty-eight days.

Likewise, the foregoing will apply if the engineer certifies in writing to the employer that in his opinion the contractor:

1 Has abandoned the contract.
2 Failed, without a reasonable excuse, to commence the works in accordance with Clause 41, or has stopped work for fourteen days after receiving written notice from the engineer to proceed.
3 Failed to remove condemned or rejected goods or materials from the site, or pulled down and replaced faulty workmanship, within fourteen days of receiving written notice to do so from the engineer.
4 Failed, despite previous written warnings by the engineer, to carry out the works diligently, or is persistently or fundamentally in breach of his contractual obligations.

If a notice of determination is given under the above circumstances, it shall be given 'as soon as is reasonably possible' after receipt of the above mentioned engineer's certification.

The employer may then complete the works himself or employ someone else, using the equipment, temporary works, goods or materials, which are deemed the property of the employer under Clauses 53 and 54, and he may sell such and use the proceeds to offset any sums due to him under the contract.

The previously mentioned notice, or an additional notice within seven days of the expiry date of the original notice, may require the contractor to assign to the employer the benefit of any agreement which the contractor may have entered into for the supply of goods or materials and/or for the execution of work for the contract. The contractor is obliged to comply with such a requirement.

As soon as is practicable after entry and expulsion, the engineer must fix, determine and certify the amount (if any) earned by, or that would reasonably accrue to, the contractor and the value of any unused goods, materials, equipment and temporary works which have been deemed the property of the employer under Clauses 53 and 54.

In the event of entry, and expulsion of the contractor under Clause 63, the employer is not liable to pay the contractor any money before the expiration of the defects completion period and, then, only the amount due after ascertainment and certification of all associated costs and expenses incurred by the employer. If the amount due to the contractor is exceeded by the costs incurred by the employer, the shortfall is recoverable as a debt.

11 Administrative procedures and site meetings

Reference ICE Clauses: 14, 20–25 and 35

Programme of works

After the award of the contract, probably the first requirement of the contractor is the need to produce a programme of the works. This requirement is covered by Clause 14, which says that the contractor must submit a programme (showing the order in which he proposes to carry out the works) to the engineer for his acceptance, within twenty-one days after the award of the contract. At the same time, the contractor must also provide a written general description for the engineer of the arrangements and methods of construction which he proposes to adopt for carrying out the works. Should the engineer reject the programme, the contractor has twenty-one days from the time of the rejection to submit a revised programme.

Subclause (2) gives the engineer twenty-one days from receipt of the contractor's programme to:

1 accept the programme in writing, or
2 reject it in writing with reasons for so doing, or
3 request further information to clarify or substantiate the programme or to satisfy the engineer as to its reasonableness having regard to the contractor's contractual obligations.

If the engineer fails to take any of the above actions within the said period, the engineer will be deemed to have accepted the submitted programme.

If the contractor fails to provide the further information within twenty-one days of receiving the engineer's request so to do (or within an extended period that the engineer may have allowed), the relevant programme will be deemed to have been rejected. If, on the other hand, the contractor provides the requested further information, the engineer has a further twenty-one days in which to accept or reject the programme.

Usually the contractor is required to submit an outline programme with his tender, which will form the basis of the programme required under Clause 14.

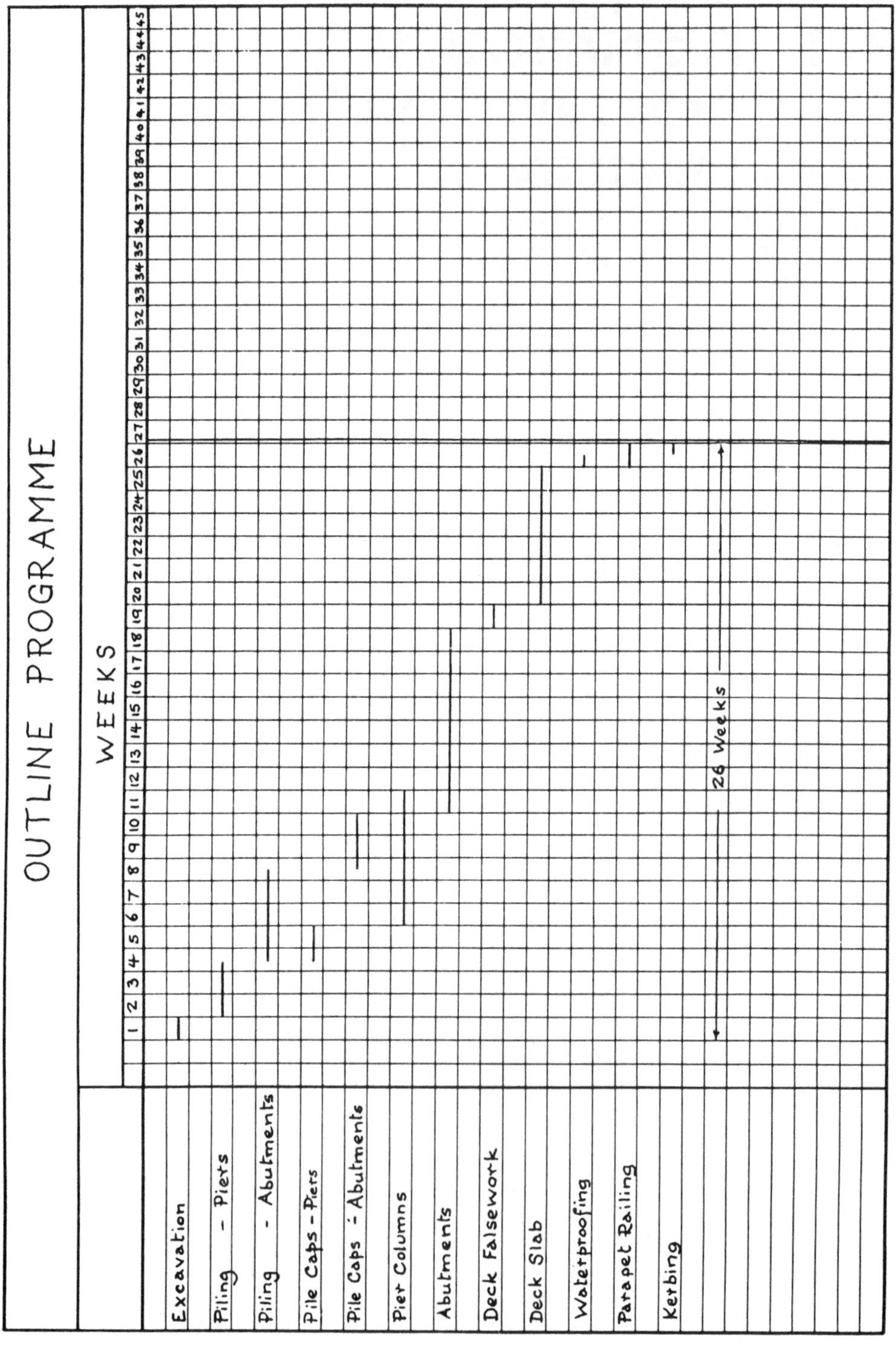

OUTLINE PROGRAMME

WEEKS

1 2 3 4 5 6 7 8 9 10 11 12 13 14 15 16 17 18 19 20 21 22 23 24 25 26 27 28 29 30 31 32 33 34 35 36 37 38 39 40 41 42 43 44 45

Excavation

Piling - Piers

Piling - Abutments

Pile Caps - Piers

Pile Caps - Abutments

Pier Columns

Abutments

Deck Falsework

Deck Slab

Waterproofing

Parapet Railing

Kerbing

26 Weeks

Figure 11.1 *Outline programme of the works for the construction of a bridge*

Such an outline programme may be in the form of a bar chart as is shown in Figure 11.1.

Where this is the case, the engineer must ensure that the component parts of the original are thoroughly broken down and itemized. This will help avoid disputes and difficulties should delay and disruption claims subsequently arise. Thus, for example, more information is likely to be required about the pile caps, piers, columns, abutment walls and deck slab in Figure 11.1. In each case, when will the following start, and how long will they take: formwork erection, reinforcement, concreting and striking formwork? In the case of the deck, details of erection and removal of temporary supports to the soffit should be provided, as should waterproofing to abutments and the deck slab.

Alternatively to the bar chart, the contractor may provide the programme in the form of a network using either arrow or precedence diagrams. Where a bar chart is employed, it can be used, additionally, during the works to monitor progress.

Subclause (4) allows the engineer to take the contractor to task at any time if the actual progress does not conform with the accepted programme. The contractor must then revise the programme to ensure completion on time, or by the later time agreed in accordance with Clause 44. In such an event, the contractor has to submit the revised programme within twenty-one days of being so instructed by the engineer.

Under Subclause (5), the engineer must provide the contractor with the necessary design criteria to enable the contractor to comply with the methods of construction and the conditions appertaining to the engineer's consent.

Should the engineer request further details of the contractor's methods of construction and related calculations, Subclause (6) requires the contractor to provide them at such times and in such detail as the engineer may reasonably require and, under Subclause (7), the engineer must inform the contractor in writing within twenty-one days after receipt of the information etc. either that he consents to the methods or in what respects they fail to meet the contract requirements or will be detrimental to the permanent works. In the latter case, the contractor must take the necessary steps, or make changes to meet the requirements and obtain the engineer's consent. Once the consent has been obtained, the contractor is not permitted to change the methods without the engineer's further consent in writing. Such further consent should not be unreasonably withheld.

Subclause (8) provides for the contractor being unavoidably delayed or incurring costs due to:

1 the engineer's unreasonably delayed consent to the proposed construction methods, or
2 the engineer's Subclause (7) requirements, or limitations imposed by any of the design criteria supplied by the engineer under Subclause (5) that could not reasonably have been foreseen by an experienced contractor at the time of tender.

Where such delay and costs occur, the engineer must take both into account in determining any extension of time for completion (Clause 44) and any payment of claims for costs incurred (Clauses 52(4) and 60). The contractor is entitled to profit in addition to his costs.

The acceptance by the engineer of the contractor's programme and the engineer's consent to the contractor's methods of construction (Subclause (9)) do not relieve the contractor of any of his contractual duties or responsibilities.

In conclusion, it is perhaps worth noting that where the contractor does not comply with the provision of programme time requirements Clause 14 provides no penalty. However, failure to provide an acceptable programme is likely to be to his disadvantage since the contract completion date will already have been fixed and non-compliance will result in an additional time constraint.

Insurance

Insurance, in a work context, falls into two principal categories: a general requirement at law under the Employers Liability Act 1969; and contractual requirements.

In the first case, all employers must be insured for claims made against them by their employees (including part-time employees) who are injured at work or contract a disease arising out of their work. Such claims may result from acts of persons other than the employer, for instance another employee, provided that the injured person was doing something that he or she was employed to do. For example, the person would be unlikely to receive compensation if the injury resulted from private work during working hours, or horseplay at work (unless that was a requirement of the job). Failure to insure in accordance with the Act is a criminal offence which could result in a heavy daily fine. Employers are required by law to:

1. Display a copy of their Employers Liability Insurance Certificate at each of their business premises, so that they can be easily read by every employee. (Failure could render the employer liable to a fine.)
2. Send the certificate, or a copy, to the Health and Safety Executive if requested so to do.
3. Produce ditto to a Health and Safety enforcement officer if so requested.
4. Allow the policy to be inspected by a Health and Safety inspector.

There are exceptions to the foregoing, whereby the employer does not have to insure close relatives, such as a spouse, son, daughter, step-child, parent, brother or sister who are his or her employees. However, regardless of any legal requirements, it would be imprudent not to insure against claims from such persons since the employer would then be unprotected if sued under common law.

The contractual requirements covering insurances of the works, persons, property and workpeople, and indemnity are to be found in Clauses 20–25 and are as follows:

Clause 20 states that the contractor is fully responsible for the *care of the works*, and for *materials, plant, and equipment* for incorporation therein. Responsibility lasts from the 'works commencement date' until the date of issue of the certificate of substantial completion, or until any outstanding work etc. which the contractor undertakes to finish in the defects correction period has been completed.

In the event of any loss or damage to the aforementioned and any loss or damage to the works occasioned by him in the course of his operations, the contractor must make full rectification at his own cost. On the other hand, the rectification of any loss or damage which arises from the 'excepted risks' (see below) will be met by the employer. Where the loss or damage arises from both a risk for which the contractor is responsible and an 'excepted risk', the engineer is required to apportion the cost of rectification between the contractor and the employer.

The 'excepted risks' for which the contractor is not liable are loss or damage due to:

1 The use or occupation of any part of the permanent works by the employer, his agents, servants or other contractors
2 Any fault, defect, error or omission in the design of the works (other than the contractor's design)
3 Riot, war, invasion, act of foreign enemies or hostilities
4 Civil war, rebellion, revolution, insurrection or military or usurped power
5 Ionizing radiations or contamination by radioactivity from any nuclear fuel, or waste arising from the combustion of nuclear materials or from nuclear explosions
6 Pressure waves caused by aircraft etc. travelling at sonic or supersonic speeds

Clause 21 deals with insurance. Under this clause, the contractor is obliged to insure the Clause 20 risks (other than the 'excepted risks') in the joint names of the contractor and the employer to their full replacement cost plus ten per cent to cover any additional costs that may arise, such as professional fees, demolition and removal of debris, for the periods previously stated. The contractor must also insure for any loss or damage arising during the defects correction period from causes occurring prior to the issue of the certificate of substantial completion. Any amounts not covered by insurance are to be borne by either the contractor or the employer in accordance with the requirements of Clause 20.

Clause 22 requires the contractor to indemnify the employer against all losses and claims whatsoever in respect of *death of or injury to persons* or *loss of or damage to property* (other than the works) which may arise out of, or in consequence of, the execution of the works or the remedying of defects. There are, however, a number of exceptions under Subclause (2) where the employer must indemnify the contractor. Both the contractor's and the employer's liability to indemnify the other party are reduced proportionately where the act or neglect can be attributed to the other party, his agents, servants or other contractors.

The insurance for the Clause 22 obligations and responsibilities is dealt with under Clause 23. Such insurance must be taken out in the joint names of the contractor and the employer and be for 'at least the amount stated in the appendix'. Furthermore, the policy must include a cross liability clause such that the insurance will apply to the contractor and to the employer as though they were separately insured. Such a clause would allow either party to claim against the other for damage to employees or property.

Accidents and injuries to workpeople are covered by Clause 24, which states that the employer is not liable for damages or compensation payable to them, unless resulting from an act or default of the employer, his agents or servants. The contractor is obliged to indemnify the employer against all such damages and compensation, and against all claims etc. in relation thereto.

Under Clause 25, the contractor must provide satisfactory evidence to the employer, *prior to* the works commencement date, that the required insurances have been taken out and, if so required, produce the policies and premium receipts for the employer's inspection. The terms of the insurances must be approved by the employer, and such approval should not be unreasonably withheld. Where there are excesses on the policies, they must be stated in the appendix.

If the contractor fails to produce satisfactory evidence of insurance as required under the contract, Subclause 25(3) allows the employer to take out the necessary insurance himself and deduct the cost of premiums etc. from monies due to the contractor, or to recover the costs as a debt.

Both the employer and the contractor are required to comply with *all* the conditions laid down in the insurance policies. If they fail to do so, each must indemnify the other against all losses and claims arising from such failure. This last sentence effectively means that if an insurance company does not meet a claim as a result of non-compliance by one party, that party will have to meet the claim as though he were the insurer.

Financial progress

Knowledge of the anticipated financial progress of a project, showing periods of the greatest expense, is likely to be needed by the employer for budgeting purposes, and by the contractor to forewarn him of periods of high financial outlay. A means of establishing the financial requirement is by examination of the contractor's programme in relation to the priced bill of quantities, and then drawing up a financial progress chart as shown in Figure 11.2. As with the bar chart, the actual progress can also be recorded to provide a graphic comparison. This is done by taking the value of appropriate items from each interim valuation and plotting them alongside the anticipated figures.

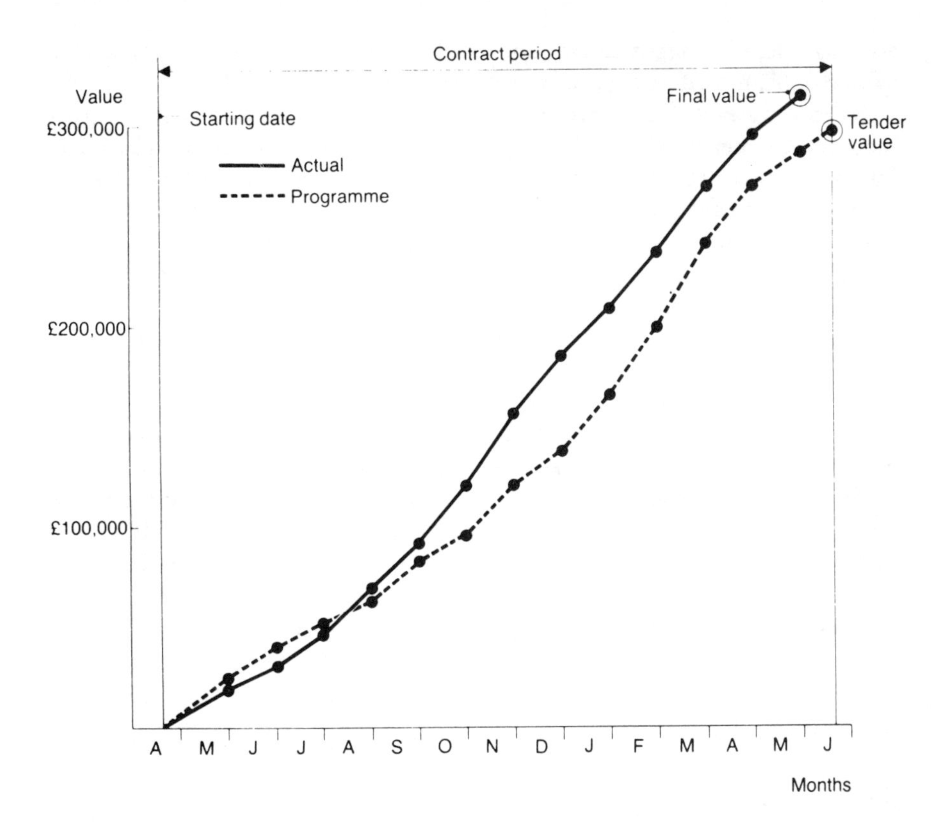

Figure 11.2 *Financial progress chart*

Site records

The successful administration of civil engineering sites depends to a large extent on good records (including charts of the types previously described), backed by a good filing system. Clearly, both the resident engineer and the contractor require systems and procedures to suit their own specific needs, but there are certain areas of mutual concern which should be well documented.

Files are required for engineer/contractor correspondence, for correspondence and arrangements made with other bodies, for example statutory undertakers, police, etc., and for dealings with the public. Registers and, in some cases, files should be kept of drawings and schedules, site instructions and variations, interim valuations, certificates, dayworks, new rates, claims and price fluctuations.

Work should be measured and, where possible, quantities agreed and finalized as the work proceeds, so that firm figures can be included in valuations. Meetings should be held frequently and at regular intervals to agree new or varied rates, and to discuss claims with a view to interim assessment.

When variations or extra works arise, they should be dealt with promptly on the spot by means of verbal or handwritten instructions to the contractor, followed shortly after by typed site instructions or variation orders. The contractor should have a set procedure for confirming verbal instructions, to avoid possible omissions on the part of the resident engineer or his staff. Likewise, there should be a standard procedure for dealing with dayworks and recording basic information.

Where additional work occurs, it should be recorded by measurement, or on a time and materials basis, and records should be agreed (even if only 'for record purposes'). This applies particularly to work to be broken out, or to be covered up. For this purpose, a useful addition to the site equipment is a polaroid camera which ensures, immediately, that a suitable pictorial record is made.

Monthly sets of progress photographs should be taken, as far as is possible, of the same subjects and from the same locations every time. On a regular basis, preferably weekly, labour and contractor's equipment returns should be produced by the contractor in accordance with Clause 35. Ideally, they should be checked and agreed before being sorted away in an appropriate file.

Adoption of the above procedures will help minimize disagreements over facts, assist in the early settlement of claims, and simplify the production and agreement of the final account.

Site meetings

Communication, co-operation and co-ordination on the site are of paramount importance with regard to effective administration of civil engineering contracts. Ideas and arrangements for putting into effect methods of construction, need to be freely discussed between the engineer's and the contractor's site staffs, so that each other's difficulties are fully understood and allowed for when making assessments and arriving at decisions. Also, where subcontractors and other parties are involved, there is a need for their problem areas to be considered, and for them to be informed of decisions made.

If no such actions are taken, or they are carried out on a casual piecemeal basis, there will be misunderstandings and confusion, which in turn may lead to hold-ups and disputes, followed by claims and delayed completion of the works. To prevent such situations arising, regular site meetings should be arranged. Depending on the size and complexity of the project, meetings may be either formal or informal.

Formal meetings

For these meetings an agenda should be drawn up covering the topics for discussion. The contents will depend on the stage that the contract work has reached and the persons invited to attend, who should be restricted to those who have something to contribute or who are directly affected by the items for discussion. Ideally, participants should receive copies of the agenda well in advance of the meeting, although this is not always practical.

Agenda for site meetings are usually set out as follows:

1 Apologies for absence.
2 Confirmation of minutes of previous meeting.
3 Matters arising from the minutes.
4 Progress of work:
   (a) Main contractor.
   (b) Subcontractors.
   (c) Statutory undertakers.
   (d) Other bodies.
5 Problems arising.
6 Information required.
7 Any other business.
8 Date and time of next meeting.

Meetings should have a chairman who is usually the engineer's assistant in a position of seniority over the resident engineer. The chairman should take charge and control the meeting, ensuring that members do not wander from the points of discussion, discuss matters privately during the meeting, or become unruly or abusive. He should allow all relevant points of view to be heard, summarize points and, where applicable, draw conclusions. Finally, he should fix the date and time of the next meeting and terminate the proceedings.

Meetings should be minuted, and copies of the minutes distributed to members before the next meeting, giving the members sufficient time to read them over and consider the contents.

Informal meetings

These meetings are likely to be held on smaller contracts, or on large contracts where relatively less important points are to be discussed. They tend to be low key affairs, with no formal chairman or agenda, and the proceedings are not minuted. As a result, they can be held at relatively short notice.

It follows that problems may arise subsequently, since there is no record or proof of what was discussed or agreed. Such meetings frequently result in misunderstandings and disputes and, sometimes, denials of what was said or agreed.

12 Public utilities, statutory undertakers and facilities for other bodies and contractors

Reference ICE Clauses: 27 and 31

Public utilities refer to services or supplies commonly available in towns or to the public generally. They are usually provided by 'statutory undertakers' or 'Statutory authorities', which have special powers and responsibilities authorized, or permitted by statute, and include the gas, electricity and water companies, and British Telecom.

Most civil engineering works, especially highway and motorway projects, include work to be carried out which affects or is carried out by statutory undertakers and other authorities, either before, during or after completion of the works. As a result, these bodies are frequently a disruptive factor in the planning and execution of the works. Contracts must be prepared, therefore, only after due regard to such concerns, and must allow adequate time for their work, and for associated works, to be carried out, since any delay on their part may have a 'knock-on' effect resulting in delay to the contract followed by disruption and delay claims from the contractor.

Statutory undertakers and public authorities

Works associated with statutory undertakers and public authorities fall into three categories:

1 Work to be carried out by such bodies themselves.
2 Contractor's work affected by their work.
3 Contractor's work which may affect their apparatus (services, pipes and cables, etc.).

In the first case, the contractor is directed to provide all reasonable facilities for other contractors, and contracts generally include separate schedules in the specification setting out the work requirements for each undertaker (see Figure 12.1). Also included are clauses directing the contractor's attention to the schedules, indicating that the work will be carried out in accordance with Clause 31 of the *ICE Conditions of Contract*, and emphasizing the need for the contractor to allow for such in his programme of works. Such clauses will additionally state that the contractor must ensure that such (stats) work does not subsequently injuriously affect the contractor's own work. If the work is carried out so as to adversely affect the contractor's own works, the contractor must inform the resident engineer.

In the second case, where the contractor's work is affected by statutory undertakers' works, the contract may also include certain works in connection with the 'stats' work to be carried out by the contractor. Where this is the case, there is an obvious need for co-operation, and co-ordination of the contractor's and statutory undertakers' works to suit their individual requirements, and to prevent hold-ups which may delay completion of the overall project.

In the third case, where the contractor's work may affect statutory undertakers' apparatus, or where the contractor is required to work near or around such, the contractor will be informed in the contract of the special requirements of each statutory undertaker, and where applicable, the lines and locations of existing services, cables and the like will be set out on the drawings, but usually it is up to the contractor to liaise with the bodies concerned to ascertain the precise locations. Preliminary meetings may be arranged with the relevant bodies prior to tender, although it is more likely that these will be deferred until after the contract has been awarded. A suitably worded clause is normally found in contracts stating that the information as to the whereabouts of existing services and mains is believed to be correct and that the information given will not relieve the contractor of his obligation under Clause 11 (inspection of site and sufficiency of tender). The clause usually continues, saying that the contractor's rates and prices must include for supporting and protecting pipes, cables and other apparatus; for keeping the engineer informed of all arrangements made with such bodies; and for ensuring that existing mains and services and the like are not interrupted without the written consent of the appropriate authority.

Should the contractor sever any existing main or service, he will normally be responsible for the damage, and for all associated costs and repairs.

Clearly, in order to assist the smooth running of the contract, early meetings with statutory undertakers and public authorities are necessary to establish details of any restrictions likely to apply to the works and, once the project is under way, there is a need for arrangements and progress to be recorded by both the contractor and the resident engineer, and for reports to be relayed to their respective head offices. Such actions will provide a history of operations should claims for delay and disruption arise at a later date.

ICE requirements

Clause 27 attempts to overcome the complex situations which arise where works occur in streets and highways, and affect various authorities and undertakings, by setting out the relationship of the contractor and the works to the Public Utilities Street Works Act 1950.

Schedule of Statutory Works		SOUTHERN ELECTRIC plc			
Drawing No.	*Diversion location*	*Details of diversion*	*Group*	*Notice required to commence the works (weeks)*	*Duration of the works (weeks)*
20 and 21	CS 2 – CS 4	Relay and joint low voltage cable north side		6 weeks	HV 13 weeks
	CS 6/7	Extend road crossing		Earliest	LV 6 weeks
	CS 3 – CS 7	Relay and joint LV cable		start	(Work can be
	CS 10	Extend road crossing		1 April 19 —	carried out
	CS 10 – CS 52	Relay and joint new HV cable(s) north side			in sections)
	CS 14	Extend road crossing			
	CS 31	Extend road crossing			
	CS 59	Extend road crossing			
		Not including street lighting cabling			
		Existing EHV cables which run the whole length of the site are to remain undisturbed			

Figure 12.1 *Example of a schedule of a statutory undertaker's works included in a specification. The right-hand columns show (a) the notice required by Southern Electric plc prior to their commencement of the works and (b) the anticipated duration of those works*

After defining the expression 'Act', the clause then requires the employer, before commencing the works, to give the contractor written notification of:

1 Whether the works, or any specific parts, are emergency works.
2 Which parts of the works are to be carried out in 'controlled' land or in a 'prospectively maintainable' highway.

Should any variations affecting the above be ordered, the employer must notify the contractor, in writing, at the time of the order. The clause points out that the employer is obliged to serve all notices required by the Act that are applicable either before, during or after completion of the works.

Except in cases of emergency works, the contractor must give a minimum of twenty-one days written notice before:

1 Commencing any part of the works in a street (as defined by the Act).
2 Commencing any part of the works in 'controlled' land or in a 'prospectively maintainable' highway.
3 Commencing as in 1 and 2 above any part of the works which is likely to affect the apparatus of any owning undertaker.

Such notices must give the date and place of commencement. Should the contractor fail to start the work within two months after the date when the notice was given, the notice is rendered invalid.

If a variation is ordered which affects a street, controlled land or prospectively maintainable highway, or emergency works results in delay and additional cost to the contractor, the engineer must make suitable allowance to the contractor by way of an extension of time and additional payment.

Except where the Act imposes requirements or obligations upon the employer, the contractor must comply with any other obligations of the Act in relation to carrying out the works, and must indemnify the employer against any liability which the employer may incur in consequence of any failure to comply.

Facilities to be provided for other contractors are covered by Clause 31, which states that the contractor must afford all reasonable facilities for parties or persons who may be working on, or near, the site of any work not included in the contract; or of any contract of the employer connected with or ancillary to the works. Such parties or persons include:

1 Any other contractors and their workmen employed by the employer.
2 Workmen of the employer.
3 Any other properly authorized authorities or statutory bodies.

Any delay suffered by the contractor, or any cost beyond that which could reasonably be foreseen by an experienced contractor at the time of tender, which results from the foregoing, must be taken into account by the engineer, and he should allow the contractor a suitable extension of time and payment of all additional reasonable costs.

13 Synopsis of the ICE Conditions of Contract

This chapter briefly summarizes clauses and subclauses of the *ICE Conditions of Contract* which have not been dealt with in earlier chapters. Where clauses have been referred to previously, the relevant chapters are indicated for easy reference.

Clause 1 Definitions

The following are defined: employer, contractor, engineer and engineer's representative. *Chapters 3 and 4*.

Also defined are the contract, specification, drawings, bills of quantities, tender total, contract price, prime cost (PC) item, provisional sum, nominated subcontractor, permanent works, temporary works, works, work commencement date, certificate of substantial completion, defects correction period, defects correction certificate, section, site, and contractor's equipment. Reference is made to singular words, which are to be read as plural, and vice versa; to headings and marginal notes, which are included to provide easy reference; to clause references; and to 'cost', which includes on or off-site overheads but no profit.

Clause 2 Engineer and engineer's representative

Includes the duties and authority of the engineer and the naming of the engineer, engineer's representative, delegation of duties, etc., appointment of assistants, instructions, reference on dissatisfaction, and impartiality. *Chapters 4 and 7*.

Clauses 3 and 4 Assignment and subcontracting

Chapters 3, 4 and 6

Clauses 5, 6 and 7 Contract documents

Includes the need for the contract documents to be mutually explanatory; the supply of documents, further drawings, specifications and instructions; notice by the contractor, delay in issue, documents to be kept on site, permanent works designed by the contractor, and responsibility unaffected by approval. *Chapter 2.*

Clause 8 The contractor's general and design responsibilities

Includes the contractor's responsibility for safety of site operations. *Chapter 4.*

Clause 9 Contract agreement

The contractor must enter into and execute a contract agreement as set out in the conditions, if called upon to do so.

Clause 10(1) Performance security

Chapter 2.

Clause 10(2) Arbitration upon security

For the purposes of arbitration in security, the employer is deemed a party to the security and any decision, award, etc. concerning the discharge date for security shall be without prejudice to the outcome of any Clause 66 settlement.

Clause 11(1) Provision and interpretation of information

The employer is deemed to have made available to the contractor, before the submission of the tender, all information relating to the ground and subsoil obtained by or on behalf of the employer, and the contractor is responsible for the interpretation of such information.

Clause 11(2) and (3) Inspection of site and sufficiency of tender

Chapter 3.

Clause 12 Adverse physical conditions and artificial obstructions

If the contractor encounters physical conditions (other than weather conditions or conditions due to weather conditions) that could not reasonably have been foreseen by an experienced contractor, the contractor must give written notice to the engineer as early as is practicable. If the contractor also intends to claim for additional payment or an extension of time he must, either then or as soon as is reasonable, inform the engineer in writing specifying the condition or obstruction to which the claim relates. The contractor is also required to give details of any of the following:

1 The anticipated effects
2 The measures he has taken, or is proposing to take
3 Their estimated cost
4 The extent of the anticipated delay in, or interference with, the execution of the works

Following receipt of such a notice, the engineer may, if he thinks fit:

1 Require the contractor to investigate and report upon the practicability, cost and timing of alternative measures
2 Give written consent to the contractor's measures, perhaps with modifications
3 Give written instructions as to how to deal with the problem
4 Order a suspension of, or variation to, the works

If the engineer decides that, either in whole or in part, the conditions or obstructions could have been reasonably foreseen by an experienced contractor, he must immediately so inform the contractor in writing.

On the other hand, where the contractor claims an extension of time or additional payment and the engineer decides that the conditions or obstructions could not reasonably have been foreseen, he must determine the amount of any costs reasonably incurred together with a reasonable percentage addition for profit, and any extension of time entitlement. The contractor must be notified accordingly with a copy to the employer. *Chapters 7 and 8.*

Clause 13 Work to be to the satisfaction of the engineer

Providing it is not legally or physically impossible, the contractor must construct and complete the works in accordance with the contract to the satisfaction of the engineer and must follow his instructions implicitly. Such instructions must come directly from the engineer or from the engineer's representative as authorized under Clause 2. The materials, plant, labour, and the mode, manner and speed of construction must be acceptable to the engineer.

If the engineer issues instructions in accordance with the foregoing or pursuant to Clause 5, involving the contractor in delay or disruption and causing him additional costs beyond those that could reasonably have been foreseen by an experienced contractor at the time of tender, the engineer must:

1 Take the delay suffered into account in determining an extension of time.
2 Subject to the contractor giving notice of a claim and keeping records in accordance with Clause 52(4), authorize payment to the contractor of reasonable costs and profit unless the delay and extra costs result from the contractor's default.

Should the instructions require a variation order, it will be deemed to have been given.

Clause 14 Programme, methods of construction, and engineer's consent

Covers the programme of works, action by the engineer, provision of further information, revision of programme, design criteria, methods of construction, engineer's consent, delay and extra cost, and responsibility. *Chapter 11.*

Clause 15 Contractor's superintendence and agent

The contractor must provide all necessary superintendence of the works with sufficient persons having adequate knowledge, for as long as the engineer considers necessary. The contractor's agent must be competent and approved in writing by the engineer, but such approval may be withdrawn at any time. The contractor's agent is to be constantly on the works, giving his whole time to the task. He is, effectively, the contractor and is in full charge of the works, receiving all instructions and directions, and being responsible for the safety of all operations.

Clause 16 Removal of contractor's employees

The contractor should employ only careful, skilled and experienced persons; and the engineer is entitled to object to any person, and require the contractor to remove that person from the works, if he is of the opinion that such person:

1 Has misconducted himself.
2 Is incompetent or negligent.
3 Has failed to conform with any safety provisions set out in the specification.
4 Has persisted in conduct prejudicial to safety or health.

Such persons cannot be re-employed upon the works without the engineer's permission.

Clause 17 Setting-out

Setting-out is the contractor's responsibility. He must bear all costs of rectifying mistakes, unless they are attributable to incorrect data supplied in writing by the engineer. The checking of any setting-out by the engineer does not relieve the contractor of his responsibilities for setting out.

Clause 18 Boreholes and exploratory excavation

Boreholes or exploratory excavation made by the contractor to the engineer's requirements must be ordered in writing as a variation, unless covered by a provisional sum or a prime cost item.

Clause 19 Safety and security, and employer's responsibilities

The contractor is responsible for the safety of all persons entitled to be on the site, and must keep that part of the site and works under his control in an orderly state. Among other things, the contractor must provide and maintain at his own cost all lights, guards, fencing, warning signs and watching when and where necessary or required by the engineer, or any competent statutory or other authority, for the protection of the works or for the safety or convenience of the public. If the employer's own workmen, or other contractors, carry out work on the site, the above responsibilities pass, as applicable, to the employer.

Clauses 20–25 Responsibilities, excepted risks and insurance

Covers care of the works, excepted risks, rectification of loss or damage; insurance of the works, and extent; damage to persons and property, exceptions, indemnity by the employer, and shared responsibilities; third party insurance, cross liability, and amount; accident or injury to workpeople; and evidence and terms of insurance, excesses, remedy on contractor's failure to insure, and compliance with policy conditions. *Chapter 11.*

Clause 26(1) and (2) Giving of notices and payment of fees

The contractor is obliged to give all notices, and to pay all fees required by Acts of Parliament, regulations, or by-laws of local or other statutory authorities relative to the execution of the works; and the employer must reimburse the contractor all sums certified by the engineer in respect of such fees, and all rates and taxes paid exclusively for the purposes of the contract.

Clause 26(3) Contractor to conform with statutes, etc.

Chapter 6.

Clause 27 Public Utilities Street Works Act 1950

Includes definitions, notifications by employer to contractor, service of notices, failure to commence street works, delays attributable to variations, and contractor to comply with other obligations of the Act. *Chapter 12.*

Clause 28 Patent rights and royalties

The contractor is required to indemnify the employer against all claims and proceedings concerned with patent rights and the like, except where the infringement results from compliance with the design or specification not provided by the contractor, in which case the employer must indemnify the contractor. The contractor is required to pay all royalties and other payments due for materials required for the works.

Clause 29 Interference with traffic and adjoining properties, noise disturbance and pollution

All operations necessary for the execution of the works must be carried out with the minimum of inconvenience to the public, as regards noise, disturbance or other pollution, and the contractor must indemnify the employer in respect of all claims relating to such matters (see also Clause 22). Conversely, the employer is required to indemnify the contractor against liability for noise and pollution which is the *unavoidable* consequence of carrying out the works.

Clause 30 Avoidance of damage to highways, and transportation of contractor's equipment and materials

The contractor must take all reasonable precautions to prevent highways and bridges communicating with or on routes to the site being subjected to extraordinary traffic within the meaning of the Highways Act 1980, or, in Scotland, the Roads (Scotland) Act 1984 and, where such traffic does arise, must select routes and vehicles and restrict and distribute loads as far as is reasonably possible to avoid damage to highways and bridges. The contractor is responsible for the necessary costs of strengthening or repairing damaged bridges, or altering or improving any highways to be used for transporting contractor's equipment or temporary works, and must indemnify the employer against all associated claims. If damage occurs to any bridge or highway arising from the transportation of materials or manufactured or fabricated articles, the contractor must notify the engineer as soon as he is aware of such or he receives a claim from the relevant authority. In certain instances, the employer may be required to pay all sums in respect of claims resulting from the transportation of materials and fabricated articles, and to indemnify the contractor in respect thereof.

Clause 31 Facilities for other contractors and delay and extra cost

Chapter 12.

Clause 32 Fossils etc.

All fossils, coins, articles of value or antiquity and structures, etc. of geological or archaeological interest discovered on the site are deemed to be the property of the employer, and reasonable precautions must be taken by the contractor to prevent their removal or damage. The contractor must also inform the engineer of any discoveries immediately, and their disposal must be as ordered by the engineer. Such orders are to be carried out at the employer's expense. In reality, the employer may find that certain articles found on the site are treasure trove owned by the Crown.

Clause 33 Clearance of site on completion

The contractor is responsible for site clearance on completion of the works. He must leave the site and permanent works clean and in a workmanlike condition to the satisfaction of the engineer.

Clause 34 Not used

Clause 35 Returns of labour and contractor's equipment

The contractor is required to supply the engineer with labour and equipment returns in the form, and at such intervals, as requested by the engineer. This obligation also applies to subcontractors.

See also *Chapter 11*.

Clauses 36–40 Workmanship and materials

Included in these clauses are the requirements regarding tests and samples, and their cost; engineer's access to site; examination of work before covering up, uncovering and making openings; removal of unsatisfactory work and materials, default of contractor in compliance, and failure to disapprove; and suspension of the work. *Chapter 6*.

Clauses 41–46 Commencement, time and delays

The items dealt with include the works commencement date, and start of works; possession of site and access, failure to give possession, and facilities provided by the contractor; time for completion; extension of time for completion, assessment of delay, interim grant of extension of time, assessment at due date for completion, and final determination of extension; night and Sunday work; and rate of progress, permission to work at night or on Sundays, and provision for accelerated completion. *Chapter 5*.

Clause 47 Liquidated damages for delay

This clause includes liquidated damages for delay in substantial completion of the whole of the works and sections of it, damages not a penalty, limitation of liquidated damages, recovery and reimbursement of liquidated damages, and intervention of variations, etc. *Chapter 8*.

Clause 48 Certificate of substantial completion

Included in this clause is notification and certification of substantial completion, premature use by the employer, substantial completion of other parts of the works, and reinstatement of ground. *Chapter 5.*

Clauses 49 and 50 Outstanding works and defects

Clause 49 requires the contractor to complete any outstanding works within the time(s) agreed with the engineer, or as soon as practicable during the defects correction period. The engineer must inspect the works before the end of the above period, and inform the contractor in writing within fourteen days of its expiry of all remedial works required.

Where the cost of remedial work is due to the use of materials, or to workmanship not in accordance with the contract, or to the contractor's neglect or failure to comply with his obligations, it must be borne by the contractor. In other cases, the work should be treated as additional work and valued accordingly. Failure by the contractor to carry out the remedial work entitles the employer to make alternative arrangements and, where appropriate, to charge the contractor with the costs. *Chapters 5 and 8.*

Clause 50 requires the contractor to carry out all necessary investigations requested in writing by the engineer to establish the cause of any defects or faults, etc. The costs must be borne by the contractor if such defects and faults are attributable to him. In other circumstances, they must be borne by the employer.

Clauses 51 and 52 Alterations, additions and omissions

These clauses cover ordered variations, variations not to affect the contract, and changes in quantities; valuation of ordered variations, engineer to fix rates, daywork, and notice of claims. *Chapter 7.*

Clauses 53 and 54 Property in materials and contractor's equipment

The clauses deal with 'vesting', that is, the transfer of ownership.

Clause 53 states that all the contractor's equipment, temporary works, goods and materials, when *on site*, are deemed to be the property of the employer, although the employer will not normally be liable for their loss or damage. The contractor must obtain the engineer's written consent before removing such items from site, but that consent must not unreasonably be withheld where the items are no longer immediately required. The clause also provides for the employer to dispose of the aforementioned items if they are left on site for an unreasonable time after completion of the works, and to retain any costs or expenses incurred from monies due to the contractor.

Clause 54 deals with the payment and vesting of goods and materials *not on site*, that is before delivery. Such articles must be listed in the appendix to the form of tender. If the contractor seeks payment, the goods and materials must be substantially ready for incorporation in the works, and must be the contractor's property. The contractor is then required to follow a set procedure, and once payment is made the articles become the employer's property although, clearly, the contractor must have possession of them for the purposes of delivery and incorporation in the works.

Upon cessation of the contractor's employment before completion of the works, the employer has the right to *any vested* goods or materials, and the contractor is required to deliver such to the employer or, failing delivery, the employer may take them from the contractor and recover from the contractor the cost of so doing.

Subcontractors must also be subject to the provisions of these clauses.

Clauses 55, 56 and 57 Measurement

Included in these clauses are quantities and correction of errors; measurement and valuation, increase or decrease of rate, attending for measurement, and daywork; and method of measurement. *Chapter 7.*

Clauses 58 and 59 Provisional and prime cost sums and nominated subcontracts

Clause 58 deals with provisional sums and prime cost items. Where the former are used, the engineer may order either or both the work, or the goods, materials or services supplied to be valued in accordance with Clause 52; or be carried out or provided by a nominated subcontractor. Where prime cost items are used, the engineer may order either or both that the contractor employ a subcontractor nominated by the engineer to execute the work, or supply the goods, materials or services; or, with the contractor's consent, that such be carried out or provided by the contractor. In this latter case the contractor is paid either in accordance with a submitted and accepted quotation or in accordance with Clause 52.

Where provisional sums and prime cost items include matters of design and specification, such requirements must be expressly stated in the contract and included in any nominated subcontract. The contractor's design and specification obligations are limited to that which is stated in accordance with Clause 54.

Clause 59(1), (2), (3) and (4) includes nominated subcontractors – objection to nomination, engineer's action upon objection to nomination or upon determination of nominated subcontract, contractor responsible for nominated subcontractors, nominated subcontractor's default, termination of subcontract, engineer's action upon termination, delay and extra expense, and reimbursement of contractor's loss. *Chapter 3* (see also *Chapter 6*).

Clause 59(5), (6) and (7) includes provisions for payment, production of vouchers etc., and payment to nominated subcontractors. *Chapter 7.*

Clauses 60 and 61 Certificates and payment

These clauses set out the procedure for monthly statements and payments, minimum amount of certificate, final account, retention, interest on overdue payments, correction and withholding of certificates, copy certificate for the contractor, and payment advice; the defects correction certificate and unfulfilled obligations. *Chapters 7 and 8.*

Clause 62 Urgent repairs

Should it be necessary for urgent remedial works to be executed to prevent accident, etc., and the contractor is unable or unwilling to carry them out immediately, the employer may make alternative arrangements for such works to be carried out. Where appropriate, the costs and expenses so incurred must be met by the contractor.

Clause 63 Determination of the contractor's employment

This clause also includes completing the works, assignment to the employer, payment after determination and valuation at date of determination. *Chapter 10.*

Clause 64 Payment in event of frustration

If the contract is frustrated by war or some other supervening event, the employer must pay the contractor all amounts due up to the date of frustration. Such amounts are based partly on rates in the contract and partly on assessments of what is fair and reasonable.

Clause 65 War clause

This is a lengthy clause setting out the procedure to be followed should there be an outbreak of war and general mobilization during the contract period. If the works can be completed within twenty-eight days of the outbreak, the contractor is obliged to continue and complete the works. If not, the employer may determine the contract.

Clause 66 Settlement of disputes

This clause also includes notice of dispute, engineer's decision, effect on contractor and employer, conciliation, arbitration, president or vice-president to act, arbitration – procedure and powers, and engineer as witness. *Chapter 9.*

Clause 67 Application to Scotland

Contract works carried out in Scotland are construed and operate as Scottish contracts subject to Scots law. *Chapter 9.*

Clause 68 Notices

Notices served on the contractor in accordance with the contract must be in writing and sent by post to, or be left at, the contractor's principal place of business or registered office. Similar arrangements are required for notices served on the employer.

Clause 69 Labour – tax fluctuations

This clause deals with various taxes etc. (including National Insurance contributions, but excluding income tax and any levy payable under the Industrial Training Act 1964), payable by, or to, the contractor or his subcontractors in respect of their workpeople engaged on the contract. It sets out the procedure for reimbursing *either party* for fluctuations that occur after the date for the return of tenders.

Clause 70 Value added tax (VAT)

This clause states that the contractor's tender is deemed not to include VAT, and that all payment certificates issued by the engineer must be net of VAT, although payments must additionally and separately identify, and the employer pay the contractor, any VAT due. Where VAT disputes between either the employer or the contractor and the Commissioners of Customs and Excise arise, each is required to render the other support and assistance as necessary to resolve the dispute. Finally, Clause 66 (settlement of disputes by arbitration etc.) is not applicable to this clause).

Clause 71 Special conditions

Any special conditions, including contract price fluctuation, which it is desired to incorporate in the conditions of contract, must be numbered consecutively starting with 71.

Glossary of terms

Acceptance: Unconditional acceptance of an offer to form a contract, usually when contracts are signed.

Ad hoc: For this purpose.

Admeasurement: Additional measurement. Measurement of contract works, as the work is carried out.

Agent: A person with authority to act for someone else.

Arbitration: The resolving of a dispute by way of a solution imposed by one or more impartial persons.

Arbitrator: One who arbitrates.

Assignment: The transfer by one party to a contract of his legal title or interest to another person.

Award: An arbitrator's finding or decision which may be:
interim, following an interim arbitration, where the works have not been completed;
final, the arbitrator's finding following the hearing;
summary, payment of a reasonable proportion of the final amount likely, in effect, an on account payment.

Bankrupt: A person who cannot pay his debts, and who is adjudicated by a court order to be a bankrupt.

Bond: Security for the proper performance of a contract, where the obligor binds himself to pay money to a specific amount, usually not exceeding 10 per cent of the tender total.

Breach: Non-fulfilment of a contractual obligation or duty.

By-laws (bye-laws): Rules, or local laws, made by subordinate authority, for the regulation of local authorities' functions, and having the full force of law.

Capacity: The legal ability to enter into a contract.

Chancery: A division of the High Court dealing principally with financial disputes, trusts and estates.

Claim: A request from a contractor for additional payment to cover work, the payment for which is not allowed for in the contract; or for an extension of time.

Common law: Law based on usage and custom. It is not written down, and does not result from legislation.

Conciliation: The bringing together of the parties in an attempt to settle their dispute. A preliminary step in the ICE arbitration procedure.
Conciliator: One who conciliates.
Condition: An important term in a contract, expressing matters basic to the contract.
Consideration: That which each party contributes to a contract.
Contingencies: Sums included in bills of quantities to cover payment for work that might arise, but is unforeseen at the time of tender.
Contract: An agreement enforceable at law.
Contract price: The amount of the final account.
County Court: Local court restricted to dealing with civil matters under set financial limits.
Court of Appeal: Court to which losing parties may take their cases following decisions in the County Court, High Court and Crown Court.
Crown Court: Court which deals principally with criminal trials.
Damages: Legal compensation paid to an injured party in a civil action.
Daywork: Additional or substituted work executed on a time recorded basis, whereby the contractor is paid his agreed costs plus one or more percentage additions for profit. Daywork may be a suitable method of payment for part of a contract, or for an entire small works contract where traditional billing is not considered suitable or where time constraints require an early start and completion of the works.
Deed: A written instrument, which must be signed, sealed and delivered, proving and testifying the agreement of the parties.
Deemed: To be treated as.
Defendant: A person sued in a civil action at law.
Deponent: A person who makes a deposition; who gives evidence under oath.
Determine: To terminate a contract.
Domestic subcontractor: A subcontractor chosen by the main contractor, as opposed to one nominated by the engineer.
Equity: A branch of common law based on fairness and conscience, which endeavours to remedy the injustices of rigorous application of the law.
Forfeiture: A contract provision whereby one party may strip the other of his interests in the contract.
Frustration: The discharge (termination) of a contract due to some intervening event beyond the control of the parties.
High Court: Court which deals with civil actions.
Indemnify: To free one from the consequences of an act.
Injunction: A court order requiring a person to do, or refrain from doing, a particular act.
Insolvency: Inability to pay debts in full.
Inter alia: Among other things.
Law Reports: Published accounts of cases giving decisions, legal proceedings and reasons for judgements.

Lien: The right to retain possession of the property of another as security until certain obligations have been met.

Limitation period: Period prescribed by statute within which actions must be brought.

Liquidated damages: A sum provided by a contract that is payable in the event of a breach. It should be based on genuine pre-estimate of the likely loss resulting from the breach, and is deemed not to be penalty.

Liquidation: The winding up of a company or partnership.

Maintenance period: The period of maintenance named in the appendix to the form of tender, calculated from the date of completion of the works (frequently twelve months). Replaced by the defects correction period in the ICE sixth edition.

Negligence: Breach of a duty of care.

Nominated subcontractor: A subcontractor chosen by the engineer (or employer), with whom the main contractor has to place a subcontract, usually for some form of specialist work.

Notice to concur: A notice served by one party on the other party in dispute requiring that party to concur in the appointment of an arbitrator.

Notice of determination: A notice served on a contractor by the employer expelling the contractor from the site and terminating the contractor's employment.

Notice of dispute: The first step in the ICE settlement of disputes procedure, whereby one party serves a notice on the engineer that requires an engineer's decision.

Notice to refer: A step in the settlement of disputes procedure whereby one party serves a notice on the other, thereby referring the dispute to arbitration.

Nuisance: Indirect interference with the use or enjoyment of another person's land.

Occupier: A person having some degree of control over the premises.

Offer: A definite promise from one party to another to be bound on specific terms: a tender.

Overheads: General costs of running a business. Also known as on-costs, indirect costs and works expenses.

Penalty: A punishment; usually, in contracts, in the form of money payments in the event of a breach.

Plaintiff: A person bringing a civil action at law.

Prime cost item: An item in a bill of quantities containing a prime cost (PC) sum to be used for the execution of work or the supply of goods, materials or services.

Principal: A person authorizing another person (an agent) to act on his behalf.

Privity of contract: The relationship existing between the parties to a contract, enabling one party to sue the other.

Provisional sum: A sum included in a bill of quantities for the execution of work or the supply of goods, materials or services which may be used at the discretion of the engineer.

Public utilities: Services or supplies commonly available in towns or to the public generally.

Quantum meruit: As much as a party providing a service has earned or deserves; a fair and reasonable price for work or goods.

Queens Bench: One of the three divisions of the High Court.

Receiving order: A court order made on presentation of a bankruptcy petition, protecting the debtor's estate.

Rectification: Correction by the court of an error in a document so that it conveys the true intention of the parties.

Repudiation: Refusal by one party to perform his contractual obligations.

Rescind: To cancel, or set aside, a contract.

Rescission: The act of rescinding.

Set-off: The amount of a counterclaim deducted from sums due or claimed to be due.

Specific performance: A court order requiring a person to carry out his contractual obligations; used where compensation is not an adequate remedy for a breach of contract.

Statute: An Act of Parliament.

Sue: To bring legal proceedings against someone in a civil action.

Surety: A person who guarantees, or undertakes responsibility for, the obligations of another person.

Stats: Abbreviation for statutory undertakers.

Statutory undertaker: A body which has special powers and responsibilities provided, or permitted, by statute.

Tender: An offer to enter into a contract for a price on stated terms.

Tender total: The total in a priced bill of quantities at the date of acceptance; the tender figure.

Tort: A civil wrong, other than in contract.

Trespass: To encroach on, or directly interfere with the possession of, another person's land or buildings, etc.

Variations: Changes to an existing contract which may be either (a) to the terms and conditions – the contract itself, or (b) to the quality and quantity of the works (variation orders).

Vesting: The transfer of ownership.

Vitiate: To invalidate (the contract).

Warranty: A minor term in a contract, subsidiary to the main purpose of the contract; a vendor's statement as to the quality of goods.

Wayleaves: Rights of way over land.

Writ (of summons): A document, issued at the instigation of the plaintiff, commencing an action against the defendant.

Index